WEATHER FOR THE MARINER

Weather for the Mariner

by William J. Kotsch, *Captain, U.S. Navy*

UNITED STATES NAVAL INSTITUTE

ANNAPOLIS, MARYLAND

Library of Congress Catalog Card Number: 79–130201
ISBN 0–87021–751–8

Preface

The aim of this book is to present in an uncomplicated and readable manner the basic principles of modern meteorology and certain practical aspects of the newly emergent science of oceanography. Intended for individuals with a love for boats, motorboats, and yachts, and with little or no previous acquaintance with the subjects of meteorology and oceanography, the book is written in a nontechnical vein. The use of calculus and high-level theory have been deliberately avoided. Instead, the book is short, nonmathematical, and contains many illustrations—to provide pleasurable as well as informative reading.

The physical phenomena of the earth's envelope of air (atmosphere) are exceedingly numerous, exciting to learn about, and of great importance to all men—particularly to those who are, in one way or another, connected with the sea. All types of small-craft or ship operations are affected to some degree by the state of the air and ocean environments—some much more so than others. Consequently, a knowledge of current weather and sea conditions *and* of forecast conditions is essential for the safe and efficient operation of all sizes of sea-going craft, in coastal waters as well as far out to sea.

Anything that disturbs the equilibrium of the earth's atmosphere is likely to produce some form of danger. The same applies to the oceans. Atmospheric storms and ocean waves, for example, force themselves on the attention of everyone who follows the sea. No seafarer can dare to ignore them. An ignorance of weather signs and sea conditions could easily be the prelude to embarrassment or failure—and perhaps to disaster. It is the author's hope that this book will provide a general knowledge of the most important weather and oceanographic "signs" and that it will help to minimize air and ocean environmental threats by use of this knowledge.

The author is extremely grateful to the U.S. Coast Guard, to members of the staff of the U.S. Naval Institute, and to the several copyright holders and many friends who have kindly made illustrations and other material available for his use.

WILLIAM J. KOTSCH
Captain, U.S. Navy

13 April 1970

FOREWORD

An appreciation of the importance of weather is universal among experienced boatmen. Our enjoyment of recreational boating is largely dependent upon the prevailing weather conditions. Furthermore, the realization that weather conditions may be extremely variable is essential if boating tragedies are to be averted.

The Coast Guard and the Coast Guard Auxiliary, recognizing these imperatives, have created the Weather Advanced Specialty Course. This text is one of a series developed by the U. S. Naval Institute for use in the Coast Guard Auxiliary Membership Training Program. It is designed to instill in the boatman an appreciation and understanding of available weather data and forecasts, but cannot and will not make professional weather forecasters of those who undertake the course. In the absence of weather forecasts, it should contribute to sound assessments of prevailing environmental conditions. The special emphasis on storms and cloud formations punctuates the necessity for being alert to sudden changes in weather conditions.

While our primary concern is the potential for adverse changes in the weather, this text will also serve to enhance the boatman's appreciation of the total boating environment, in good weather as well as bad. An understanding of the weather will add enjoyment to your days of fair sailing, and stand you in good stead when threatening weather lies over the horizon.

W. J. SMITH
Admiral, U. S. Coast Guard
Commandant

Contents

WEATHER FOR THE MARINER

Chapter One

Weather Warnings and Displays

Almost 300 years ago, Jonathan Swift wrote with great conviction, "How is it possible to expect mankind to take advice when they will not so much as heed warnings?" Admittedly, it requires more wisdom to profit from good advice than to give it. But things have changed considerably during the past three centuries—especially from the standpoint of weather and sea condition forecasts, advisories, and warnings.

Today, no self-respecting jet-age aviator would dream of taking off without a thorough knowledge of, and the latest available advisories regarding altimeter settings, icing conditions, clear-air turbulence, the jet stream, and so forth. Similarly, no self-respecting mariner at any echelon should venture from port without an understanding of, and the latest available information concerning the various types and categories of weather and wind warnings and displays which are especially designed, issued, and exhibited to keep him out of serious difficulty.

YARDSTICKS FOR WARNINGS

The most meaningful and universally accepted basis for describing and classifying the various types of meteorological warnings in coastal waters and at sea is that of *wind speed*. These warnings may be associated with middle- or high-latitude weather systems, with closed cyclonic (counterclockwise in the northern hemisphere) circulations of tropical origin inside or outside the tropics, or with weather systems of tropical origin other than closed cyclonic circulations.

Wind warnings associated with weather systems located *in latitudes outside the tropics*, or with weather systems of tropical origin other than closed cyclonic (rotary) circulations, are expressed in the following way:

WARNING TERMS	EQUIVALENT WIND SPEEDS
Small Craft Warning	Winds up to 33 knots (38 mph), and used mostly in coastal and inland waters
Gale Warning	Winds of 34–47 knots (39–54 mph)
Storm Warning	Winds of 48 knots (55 mph) or greater

Wind warnings associated with *closed cyclonic* (rotary) *circulations of tropical origin* are expressed in the following way:

WARNING TERMS	EQUIVALENT WIND SPEEDS
Tropical Depression	Winds up to 33 knots (38 mph)
Tropical Storm	Winds of 34–63 knots (39–73 mph)
Hurricane/Typhoon	Winds of 64 knots (74 mph) or greater

An easy-to-memorize system of pennants, flags and lights (for nighttime), as shown in figure 1–1, is displayed at most coastal points along the seacoasts of the United States when winds dangerous to navigation are predicted for any coastal area. This warning display scheme is as follows:

Small-Craft Warning—One red pennant displayed by day and a red light over a white light at night indicate that winds up to 33 knots (38 mph) or sea conditions dangerous to small craft operation are predicted for the area.

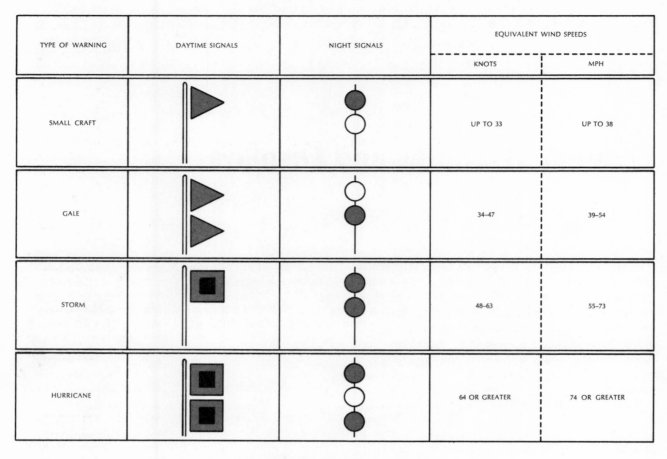

TYPE OF WARNING	DAYTIME SIGNALS	NIGHT SIGNALS	EQUIVALENT WIND SPEEDS	
			KNOTS	MPH
SMALL CRAFT			UP TO 33	UP TO 38
GALE			34–47	39–54
STORM			48–63	55–73
HURRICANE			64 OR GREATER	74 OR GREATER

Figure 1–1. Warning displays

Gale Warning—Two red pennants by day and a white light above a red light at night indicate that winds of 34–47 knots (39–54 mph) are forecast for the area.

Storm Warning—A single square red flag with a black center displayed by day and two red lights at night indicate that winds of 48 knots (55 mph) and above are predicted for the area. If the winds are associated with a tropical cyclone (hurricane), storm warnings indicate forecast winds of 48–63 knots (55–73 mph).

Hurricane Warning—Two square red flags with black centers displayed by day and a white light between two red lights at night indicate that winds of 64 knots (74 mph) or higher are forecast for the area. (Displayed only in connection with a hurricane.)

Any individual worth his salt who follows the sea is concerned for the safety and well-being of his passengers and crew as well as himself. It goes without saying that one must be thoroughly familiar with the various warning criteria and what they mean, and one must recognize instantly—*and heed* —the four types of warning displays.

For over 160 years, wind force has been conveniently expressed by means of the *Beaufort Scale,* a numerical scale devised in 1808 by Admiral Sir Francis Beaufort of the British Navy. It was based originally on the amount of canvas that a man-of-war of the period could carry, with different winds. The numbers on the scale ranged from 0 (representing calm conditions) to 12 (representing a hurricane "such that no canvas could withstand"). With the disappearance of sailing ships, the scale as originally designed became unsuitable and has been revised on several occasions.

Table 1–1 illustrates the Beaufort Scale and enables one to estimate the Beaufort force (or wind speed) from the appearance of the sea. Beaufort number 5, for example, with winds of 17–21 knots (a knot is 1.15 miles per hour), is a fresh breeze with an effect on the sea that produces moderate waves, which take a more pronounced long form; many white horses are formed, and there is a chance of some spray.

Table 1–1   The Beaufort Wind Scale, Speed Conversions and Descriptions

Beaufort Number	Knots	Miles Per Hour	Description	Effect at sea	Wind Symbols on Weather Maps
0	0–0.9	0–0.9	Calm	Sea like a mirror.	Calm
1	1–3	1–3	Light air	Scale-like ripples form, but without foam crests.	Almost Calm
2	4–6	4–7	Light breeze	Small wavelets, short but more pronounced. Crests have a glassy appearance and do not break.	5 Knots
3	7–10	8–12	Gentle breeze	Large wavelets. Crests begin to break. Foam has glassy appearance. Perhaps scattered white horses.	10 Knots
4	11–16	13–18	Moderate breeze	Small waves, becoming longer. Fairly frequent white horses.	15 Knots
5	17–21	19–24	Fresh breeze	Moderate waves, taking a more pronounced long form. Many white horses are formed. Chance of some spray.	20 Knots
6	22–27	25–31	Strong breeze	Large waves begin to form. White foam crests are more extensive everywhere. Some spray.	25 Knots
7	28–33	32–38	Moderate gale	Sea heaps up and white foam from breaking waves begins to be blown in streaks along the direction of the wind. Spindrift begins.	30 Knots
8	34–40	39–46	Fresh gale	Moderately high waves of greater length. Edges of crests break into spindrift. Foam is blown in well-marked streaks along the direction of the wind.	35 Knots
9	41–47	47–54	Strong gale	High waves. Dense streaks of foam along the direction of the wind. Sea begins to roll. Spray may affect visibility.	45 Knots
10	48–55	55–63	Whole gale and/or Storm	Very high waves with long overhanging crests. The resulting foam in great patches is blown in dense white streaks along the direction of the wind. On the whole, the surface of the sea takes a white appearance. The rolling of the sea becomes heavy and shocklike. Visibility is affected.	50 Knots
11	56–63	64–73	Storm and/or Violent storm	Exceptionally high waves. Small- and medium-sized vessels might for a long time be lost to view behind the waves. The sea is completely covered with long white patches of foam lying along the direction of the wind. Everywhere, the edges of the wave crests are blown into froth. Visibility seriously affected.	60 Knots
12	64 or higher	74 or higher	Hurricane & Typhoon	The air is filled with foam and spray. Sea is completely white with driving spray. Visibility is very seriously affected.	75 Knots

Today, by international agreement, all wind reports are encoded and plotted on weather maps in knots. Frequent reference however is still made to the Beaufort Scale. The right-hand column of table 1–1 shows how wind reports are plotted on weather maps to the nearest five knots. Each short barb represents a speed of five knots. Each long barb represents 10 knots, and each pennant represents 50 knots. All arrows "fly with the wind"; they point in the direction toward which the wind is blowing.

BASIC DEFINITIONS

About the middle of the nineteenth century, one of England's most famous Prime Ministers, Benjamin Disraeli, stated unequivocally, "I hate definitions!" And he meant it. However, marine activity of any type is antithetical to this kind of attitude. With regard to advisories and warnings relating to weather systems of tropical origin, one must understand the terminology and catch phrases used by meteorologists, commentators, and newscasters on radio and television programs, facsimile and radio broadcasts, and automatic telephone answering services in order to realize the significance of the information or the nature of the threat and to take advantage of the warning information. The following few definitions should be at the fingertips of the reader:

Cyclone—A closed atmospheric circulation rotating counterclockwise in the Northern Hemisphere. (See figure 1–2.)

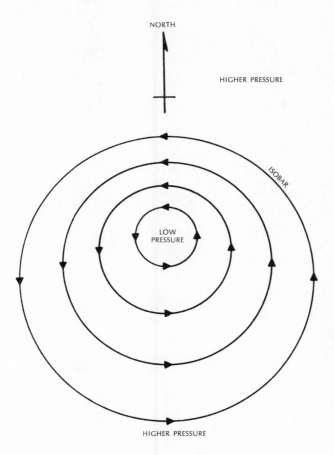

Figure 1–2.   Cyclone

Tropical Cyclone—A warm core (center warmer than the surrounding air), nonfrontal cyclone of synoptic scale (wave length of approximately 550–1350 nautical miles), developing over tropical or subtropical waters and having a definite organized circulation. (See figure 1–3.)

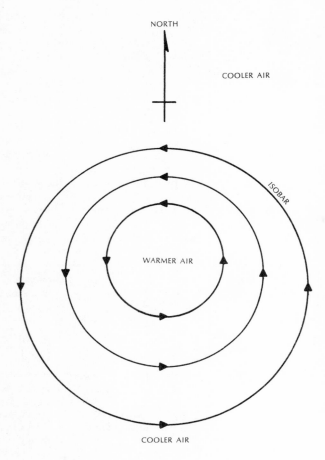

Figure 1–3.   Tropical cyclone

Tropical Disturbance—The weakest recognizable stage of a tropical cyclone in which rotary circulation is slight or absent at the earth's surface, but is possibly better developed at higher levels in the atmosphere. There may be one closed surface isobar (line of constant pressure on a weather map) or none at all, and no strong winds. (See figure 1–4.)

Tropical Depression—The weak state of a tropical cyclone with a definite closed circulation at the earth's surface, and one or more closed surface isobars, with wind speeds up to 33 knots (38 mph). (See figure 1–5.)

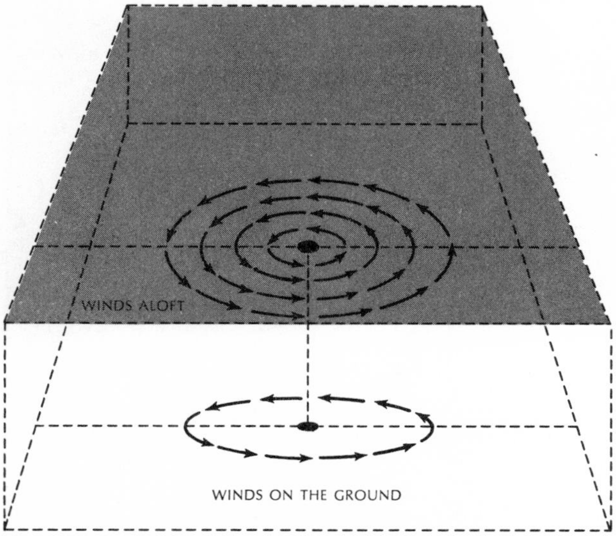

Tropical Storm—A tropical cyclone with closed isobars and highest wind speeds of 34–63 knots (39–73 mph), inclusive. (See figure 1–6.)

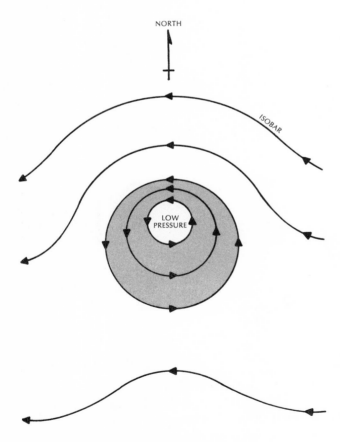

Figure 1–4. Tropical disturbance

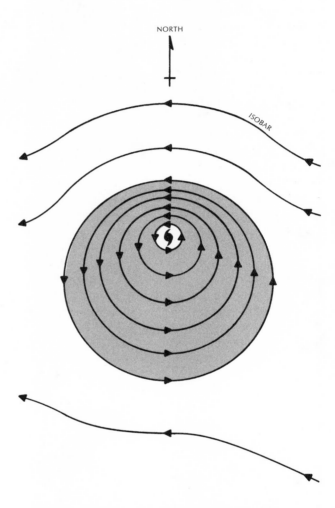

Figure 1–6. Tropical storm with highest wind speeds from 34 to 63 knots (39 to 73 mph)

Hurricane—A tropical cyclone with sustained surface winds of 64 knots (74 mph) or higher. (See figure 1–7.)

Typhoon—A tropical cyclone (in western Pacific, west of 180 degrees longitude) with sustained surface winds of 64 knots (74 mph) or higher.

Hurricane Season—That portion of the year having a relatively high incidence of hurricanes. In the North Atlantic it is usually regarded as the period from June through November, and in the east and north Pacific it is usually regarded as the period June through October.

Figure 1–5. Tropical depression with highest wind speeds up to 33 knots (38 mph)

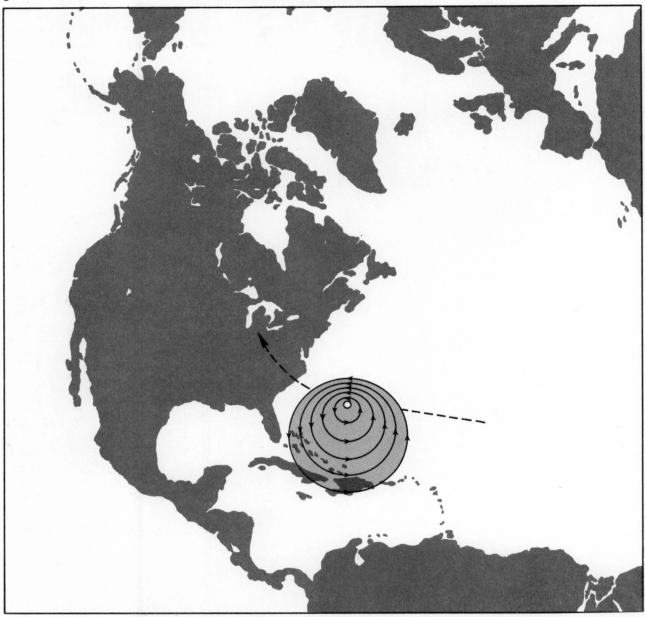

Figure 1–7.   Hurricane with sustained surface winds of 64 knots (74 mph) or higher

Typhoon Season—There is no true typhoon season. Typhoons in the western Pacific can—and do—occur in every month of the year. However, 90 percent of the typhoons occur between early June and late December. A maximum (22.6 percent) of the total occurs in August, and a minimum (0.6 percent) in February.

Quadrant—The 90-degree sector of the storm centered on a designated cardinal point of the compass. An eight-point compass rose is used when referring to quadrants.

Example: The north quadrant refers to the sector of the storm from 315 degrees through 360 degrees to 045 degrees. (See figure 1–8.)

Semicircle—The 180-degree sector of the storm centered on the designated cardinal point of the compass. A four-point compass rose is used when referring to a semicircle.

Example: The south semicircle refers to the segment of the storm from 090 degrees through 180 degrees to 270 degrees. (See figure 1–8.)

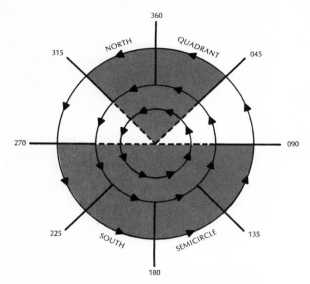

Figure 1–8. Storm quadrants and semicircles

BROADCASTS

The Environmental Science Services Administration (ESSA) was established within the U.S. Department of Commerce by Presidential order in July 1965. The formation of ESSA brought together the functions of the Coast and Geodetic Survey and the Weather Bureau, which became the two major elements of the new agency with the most immediate interest to those who follow the sea.

The Weather Bureau provides reporting, forecasting, and warning services designed to promote navigation safety among both private and commercial seamen. Weather maps, warning display signals, Coastal Warning Facilities Charts, and weather broadcasts by commercial and marine radio stations are the important means used to provide these services.

Weather forecasts for boating areas in the United States (including Alaska, Hawaii, and the Great Lakes) and in Puerto Rico are issued every six hours by ESSA's Weather Bureau. Each forecast covers a specific coastal water area—for example, "Eastport, Maine to Block Island, Rhode Island." If strong winds or sea conditions hazardous to small-boat operations are expected, the forecasts include a statement as to the type of warning issued and the area where warning signals are displayed. The various types of warnings issued and the displays for each were described in a previous section of this chapter.

Similar forecasts and warnings are issued for numerous inland lakes, dams, reservoirs, and river waterways throughout the country. Daily advices indicating expected stream flow, river gauge heights, and flood warnings, as required, are also issued by the Weather Bureau.

There is also a special announcement issued by the Weather Bureau via press and radio and television broadcasts whenever a tropical storm or hurricane becomes a potential threat to a coastal area. This is termed a *Hurricane Watch* announcement, and it is *not* a warning. The Hurricane Watch announcement indicates that a hurricane is near enough so that everyone in the area covered by the Watch should pay attention, listen for subsequent advisories, and be ready to take precautionary action *in case* hurricane warnings are issued in the near future. No visual display is provided for Hurricane Watch announcements.

Latest forecasts of all types issued are available on AM and FM radio, television, and marine radiotelephone station broadcasts. Radio stations in cities located along principal rivers include stream flow and river data in their weather broadcasts. Whenever storm and flood warnings are issued by the Weather Bureau, all stations make frequent broadcasts of these advices as a service to small-craft operators, the general public, and other interests. All advices can also be obtained by telephone from the local ESSA Weather Bureau office. Radio broadcast schedules of stations, Weather Bureau office telephone numbers, and the locations of warning display stations are as shown on the *Coastal Warning Facilities Charts.*

The Weather Bureau issues annually a series of *Coastal Warning Facilities Chart* editions for the following areas:

- Eastport, Me. to Montauk Point, N.Y.
- Montauk Point, N.Y. to Manasquan, N.J.
- Manasquan, N.J. to Hatteras, N.C.
- Hatteras, N.C. to Brunswick, Ga.
- Eastern Florida
- Apalachicola, Fla. to Morgan City, La.
- Morgan City, La. to Brownsville, Tex.
- Point Conception, Calif. to Mexican Border
- Eureka, Calif. to Point Conception, Calif.
- Canadian Border to Eureka, Calif. and Alaska
- Great Lakes: Michigan and Superior
- Great Lakes: Erie, Huron, and Ontario
- Hawaiian Islands
- Puerto Rico and Virgin Islands

Copies of these charts can be obtained from the Superintendent of Documents, U.S. Government Printing Office, Washington, D.C.

Also, the *Notice to Mariners* is prepared jointly by the U.S. Coast Guard, the U.S. Naval Oceanographic Office, and the Coast and Geodetic Survey which is issued free to those who desire copies regularly. This weekly pamphlet is issued as a safety aid to keep mariners advised of changes to charts and related publications. *Local Notices to Mariners* issued by each U.S. Coast Guard district should be used instead of the weekly *Notice* by mariners operating within one Coast Guard district.

On a much larger (worldwide) scale, weather and oceanographic information, forecasts, and warnings are also readily available. Ships and land stations, regardless of location, observe and record weather conditions simultaneously every six hours as a matter of routine, starting at midnight Greenwich time. Under certain conditions (such as a storm, hurricane, or unusual weather phenomena), weather and sea observations and reports are made at more frequent intervals. These reports are transmitted by landline or radio to designated collection and control centers and from there are relayed further so as to reach all users in the shortest possible time. These reports are then used in the preparation of weather charts, from which forecasts are made and advisories or warnings issued. The forecasts and warnings are transmitted on assigned frequencies and at specified times to any recipient who can receive the broadcast. In addition to forecasts and warnings, a general description of the weather map is also broadcast, along with selected surface weather reports from which individuals aboard ship may prepare their own weather maps, if they so desire. Advancements in radio facsimile techniques have made possible the broadcast of pictures (facsimiles) of current and predicted (prognostic) weather charts. Innumerable merchant and naval ships of many countries are equipped to copy this form of radio transmission, thereby saving many man-hours of weather plotting and analysis.

Complete weather broadcasts and their schedules for various parts of the world are contained in the following publications:

- *H. O. 118 (Radio Weather Aids)*
  Atlantic, Mediterranean, Pacific, and Indian Ocean Areas
  Published by the U.S. Naval Oceanographic Office
  Washington, D.C. 20390

- *Weather Service for Merchant Shipping and Coastal Warning and Facilities Charts*
  Published by the ESSA Weather Bureau
  Washington Science Center, Rockville, Maryland 20852

- *Information for Shipping*
  Publication No. 9, TP–4 (Volume D)
  Published by the World Meteorological Organization
  Case Postale No. 7
  CH 1211 Geneva 20, Switzerland

THE HURRICANE- AND TYPHOON-WARNING SYSTEMS

Except for the tornado, the hurricane or typhoon is the most violent and the most destructive meteorological phenomenon experienced by man—appropriately termed "nature on the rampage." These beasts of nature pack a fantastic punch. But this is not surprising when one considers the tremendous amounts of energy involved. In a single day, the average hurricane or typhoon releases an amount of heat which, if converted to electrical energy, could provide a six months' supply of electricity for the entire United States. In a single day, even a small hurricane or typhoon will release about 20 billion tons of water—an energy equivalent of almost 500,000 atomic bombs (or almost six atomic bombs per second)!

All sorts of interesting and awesome energy-equivalent calculations have been made by scientists, and all point in the same direction—hurricanes and typhoons (and their effects) are to be avoided, if at all possible.

One might logically ask, "Since there are only about five to eight hurricanes a year in the Atlantic, three to six hurricanes in the eastern and central Pacific, and 19 to 25 typhoons a year in the western Pacific, why all the hullabaloo about hurricanes and typhoons?" The answer is quite simple and perhaps, for some, astonishing.

Even though these tropical cyclones are *relatively* rare occurrences, they exact a tremendous annual toll in human life, in property damage, and in money. Since the turn of the present century, more than 20,000 Americans have lost their lives to hurricanes. And it is likely that the annual hurricane damage in the United States, alone, will average more than $100 million. Once in every ten or fifteen years, a *single* tropical cyclone will cause property damage in excess of $1 billion!

Killer hurricanes and typhoons can occur in almost any year, and it is impossible to predict well in advance, with accuracy, when these devastating phenomena will develop. Hurricane *Flora,* in 1963, was the second worst storm in the Caribbean and western Atlantic since Columbus discovered the New World. An estimated 7,200 lives were lost in Haiti, Cuba, and the Dominican Republic. In October 1962, a typhoon crossing southern Thailand claimed almost 5,000 lives. And as recently as 1967, a single tropical cyclone in the Indian Ocean was responsible for the loss of over 3,500 lives.

This is why there is so much hullabaloo about these meteorological hazards and why the threat which they constitute is not to be minimized.

Like the old Navy adage, "New broom; new long splices," hurricanes in various parts of the world are known by different names. But they are all the same basic tropical cyclone weather phenomenon:

AREA	NAME
Atlantic	Hurricane
Caribbean Sea	Hurricane
Gulf of Mexico	Hurricane
Eastern Pacific	Hurricane
Central Pacific	Hurricane
Mexico	Cordonazo
Haiti	Taino
Western Pacific	Typhoon
Philippines	Baguio *or* Baruio
Indian Ocean	Cyclone
Australia	Willy-Willy

The earliest mention of the type of storm we now call *hurricane* appeared in the voyage logs of Christopher Columbus, but he used the same terminology to describe the severe storms of both summer and winter, which are really quite different weather phenomena. The winter storms have a cold core (or center), whereas hurricanes (or any tropical vortex) have a warm core. It is generally believed that Columbus first encountered a hurricane in the vicinity of Santo Domingo in October, 1495. Some sources mention a hurricane in about the same geographical area in 1494. And in 1502, Columbus wrote about another hurricane—this one also in the vicinity of Santo Domingo. For a wealth of historical information on these very early hurri-

canes, the reader is referred to a wonderful book titled simply *Hurricane,* by Marjory Stoneman Douglas (New York: Rinehart and Co.), published in 1958.

Through the years, real progress in hurricane and typhoon forecasting has been a very slow and gradual affair, beset with many difficulties. To begin with, these storms are born and spend most of their lives at sea. Until the advent of radio, it was next to impossible to obtain observational weather data and reports in oceanic and tropical areas. In their excellent book titled *Atlantic Hurricanes,* published in 1960, Gordon E. Dunn (then Director, National Hurricane Center) and Banner I. Miller (Research Meteorologist, National Hurricane Center) report the following:

> Throughout the first decade of the present century, [hurricane] detection was a most difficult problem. When a message was received by cable indicating the presence of a hurricane, warnings or a hurricane alert were frequently issued for extensive portions of the United States coast line. On one occasion during the first years of the twentieth century a message was received in Washington, D.C., indicating that a hurricane was approaching the Antilles. During the ensuing week, hurricane warnings were displayed at one time or another from Charleston, South Carolina, to Brownsville, Texas, only to have the hurricane eventually show up at Bermuda. As late as 1909 hurricane warnings flew from Mobile, Alabama, to Charleston, South Carolina, in connection with a hurricane which eventually affected only extreme southern Florida. Not until about 1920 did forecasters begin to make a serious attempt to try to actually forecast hurricane paths.*

Fortunately for all mariners (and coastal and island residents, as well), the highly suspect, shot-gun-type hurricane and typhoon forecasts and warnings are a thing of the past. From very humble beginnings, the U.S. Weather Bureau, the U.S. Navy, and more recently the U.S. Air Force, have evolved a highly effective surveillance system of the hurricane areas of the North Atlantic Ocean (including the Gulf of Mexico and the Caribbean Sea) and the eastern and central North Pacific Ocean areas. The same type of highly effective surveillance is conducted by the U.S. Navy and the U.S. Air Force in the typhoon areas of the western North Pacific Ocean areas. Routine and special weather reports from land stations, ships at sea, and aircraft; weather satellite reports; radar reports from land stations, ships at sea, and aircraft; and the

---

* G. E. Dunn and B. I. Miller, *Atlantic Hurricanes* (Baton Rouge: Louisiana State University Press, 1960), p. 137.

specially instrumented weather reconnaissance aircraft of the Weather Bureau, Navy, and Air Force all combine to permit the pinpoint-location, detection, and tracking of hurricanes and typhoons with phenomenal accuracy. As one would expect, international cooperation in this endeavor is, for the most part, excellent.

To provide timely and accurate information and warnings regarding hurricanes and typhoons, the North Atlantic and North Pacific Oceans have been divided into geographical areas of responsibility as shown in figure 1–9. The stars show the locations of the Navy's Fleet Weather Centrals engaged in this work at Rota, Spain; at Jacksonville, Florida; at Alameda, California; at Pearl Harbor, Hawaii; and on the island of Guam, Mariana Islands. Those major Hurricane Warning Offices of the Weather Bureau shown by circles in figure 1–9 are located at Honolulu, Hawaii; at San Francisco, California;

and at the National Hurricane Center located at Miami, Florida. Not shown on the chart are the additional Weather Bureau Hurricane Warning Offices located at San Juan, Puerto Rico; at New Orleans, Louisiana; at Washington, D.C.; and at Boston, Massachusetts. While the responsibilities of the ESSA Weather Bureau are to the general public and other interests, those of the Navy and Air Force are to meet the diverse and unique requirements of the U.S. military establishment. However, extremely thorough and complete coordination in all facets of this work is effected amongst the civilian and military meteorological organizations. Details of hurricane- and typhoon-warning systems will be discussed in the following sections of this chapter.

The Hurricane-Warning System

As shown in figure 1–9, the responsibilities for issuing tropical cyclone information and hurricane

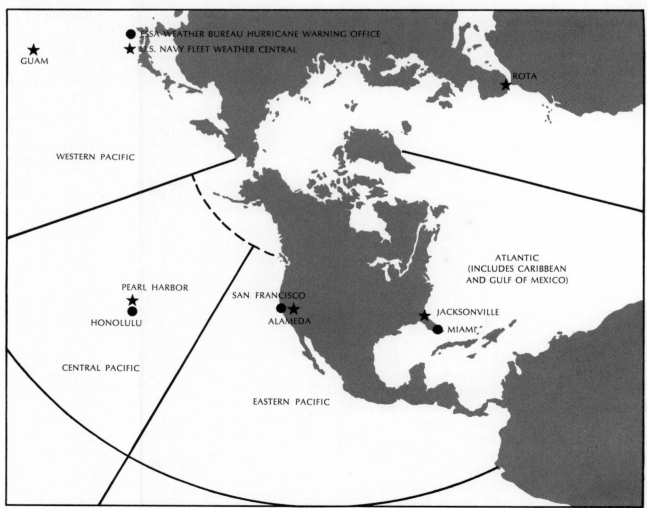

Figure 1–9. Geographical areas of responsibility of the major ESSA Weather Bureau and Navy Hurricane/Typhoon Warning Centers (Atlantic and Pacific)

warnings for interests in the eastern North Atlantic (north of the equator and east of 35 degrees west longitude) rest with the Navy's U.S. Fleet Weather Central, Rota, Spain. The Fleet Weather Central at Rota is one of the most modern, best-equipped, and busiest weather commands in the world. In addition to all the standard meteorological and oceanographic equipment and capabilities, the Fleet Weather Central, Rota has the equipment necessary to receive, record, and transmit weather satellite data and information and also has an electronic computer laboratory capable of communications at a speed of over 4,000 words per minute.

Tropical cyclone and hurricane aerial reconnaissance east of 35 degrees west longitude is conducted by the Navy's specially instrumented aircraft of Weather Reconnaissance Squadron Four, which are homeported at the Naval Air Station, Jacksonville, Florida. When required, these Super Constellation hurricane-hunter aircraft are deployed to the Azores Islands, to Rota, Spain, or to a military base in England to conduct the necessary reconnaissance flights under the direction of the Fleet Weather Central, Rota.

When a tropical storm or hurricane crosses 35 degrees west longitude traveling from west to east, the Weather Bureau ceases to issue formal public advisories. However, the Weather Bureau continues to issue marine bulletins on these dangerous weather phenomena after they pass eastward of 35 degrees west longitude, as long as they are of importance, or constitute a threat, to merchant shipping and other interests. These bulletins are included in Weather Bureau Marine Bulletins broadcast to ships four times daily at 0400, 1000, 1600, and 2200 Greenwich time over radio station NSS, Washington, D.C. Similarly, the Weather Bureau issues bulletins on tropical storms and hurricanes in progress when they are east of longitude 35 degrees west in the North Atlantic, but are moving westward. These bulletins are included in the Weather Bureau shipping bulletins broadcast to merchant ships via radio station NSS.

When a tropical storm or hurricane is traveling westward and crosses 35 degrees west longitude, responsibilities for the continued issuances of forecasts and warnings to the general public, shipping interests, etc., are transferred to the world-famous National Hurricane Center, Miami. At the same time, responsibilities for the issuance of forecasts and warnings to the U.S. military establishment are transferred from the Fleet Weather Central, Rota, to the Navy's Fleet Weather Facility, Jacksonville. Aircraft reconnaissance of tropical storms

and hurricanes in the western North Atlantic (west of 35 degrees west longitude), the Gulf of Mexico, and the Caribbean Sea, is conducted by the superconnies of the Navy, the WC–130 (Hercules) and WB–47 aircraft of the Air Force, and the weather reconnaissance aircraft of the ESSA Weather Bureau Research Flight Facility.

The ESSA Weather Bureau hurricane-forecast centers and their respective districts in the eastern and southern United States and adjacent water areas are: *Boston* for the New England coast and adjacent waters; *Washington* for coastal and marine areas from Cape Hatteras, North Carolina, north to New York and Long Island; *Miami* for the western Caribbean, the extreme eastern Gulf of Mexico, the Atlantic coast south of Cape Hatteras, and the Atlantic Ocean south of latitude 35 degrees north and eastward as far as weather reports are available; *San Juan* for the Caribbean Sea and the islands east of longitude 75 degrees west and south of latitude 20 degrees north; and *New Orleans* for the western and central Gulf of Mexico.

Referring again to figure 1–9, in the eastern Pacific (east of 140 degrees west longitude), responsibility for the issuance of tropical storm and hurricane advisories and warnings for the general public, merchant shipping, and other interests rests with the Weather Bureau Hurricane Warning Office, San Francisco. For the U.S. military establishment, it rests with the Navy's Fleet Weather Central, Alameda. Formal advisories and warnings are issued at 0300, 0900, 1500, and 2100 Greenwich time, and are included in the marine bulletins broadcast by radio stations KPH, KMI, KFS, NMC, ELH, KOE, KOK, NMQ, and KOU.

In the central Pacific (between the 180th meridian and 140 degrees west longitude), the responsibility falls to the Weather Bureau Hurricane Warning Office, Honolulu for civilian and related interests, and to the Navy's Fleet Weather Central, Pearl Harbor for the U.S. military establishment. Formal tropical storm and hurricane advisories and warnings are issued at 0300, 0900, 1500, and 2100 Greenwich time, and are included in the marine bulletins broadcast by Radio Station KHK (Kahuka, Hawaii). Broadcasts are made every two hours on the even half-hour during the hours of station operation.

Tropical storm and hurricane aircraft reconnaissance in both the eastern and central North Pacific Ocean areas is conducted by the U.S. Navy and the U.S. Air Force.

The hurricane-warning service of the United States has three principal functions: (1) the collection of the necessary observational data, (2) the preparation of timely and accurate forecasts and warnings, and (3) the rapid and efficient distribution of advices, warnings, and all other pertinent hurricane information to all civilian and military consumers.

Warnings and advisories are prepared and distributed, or broadcast, by the various centers and centrals as soon as a tropical storm or a hurricane is discovered and for as long as it remains a hurricane or a threat to life, property, ships, and smaller craft. The messages contain the position of the storm/hurricane, its intensity, its direction and speed of movement, and a description of the area of strong winds. Also included is a forecast of future movement and intensity. When the storm center is over the open sea, cautionary advices are given ships and small craft; and when it is approaching or likely to affect any land area, details on when and where it will be felt and data on tides, floods, and maximum wind are included.

Normally during the progress of a hurricane, advisories and warnings along the U.S. eastern seaboard and adjacent maritime areas are issued at 0400, 1000, 1600, and 2200 Greenwich time (2300, 0500, 1100, and 1700 EST). Occasionally, rapid developments or the receipt of later reports will require the issuance of interim advisories. After hurricane warnings have been issued for some section of the coast line, bulletins may be issued as often as once every hour, or as often as required.

Widespread and rapid distribution is given these warnings, advisories, and bulletins by all available means of communications, including the hurricane teletype systems and electronic computer data-links which make them immediately available to all

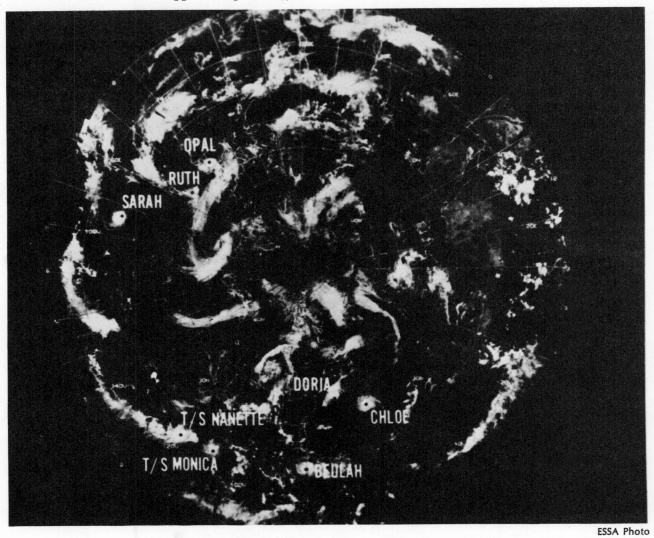

ESSA Photo

Figure 1–10. Eight major tropical storms as seen by a weather satellite

meteorologists almost simultaneously. Each local weather office then gives the warnings intensified distribution by radio, telephone, television, newspaper, and other means in its area of responsibility. Arrangements are made with local radio stations for immediate broadcast, often from microphones located in the weather offices. All of these hurricane advices are also immediately sent from the forecast centers to the press associations, which give them the highest priority distribution.

The Typhoon-Warning System

Situated atop Nimitz Hill on the western side of the beautiful tropical island of Guam in the western North Pacific (near latitude 14 degrees north, longitude 145 degrees east), the Navy's Fleet Weather Central, Guam, with its built-in Joint (Navy and Air Force) Typhoon Warning Center (JTWC), is responsible for all tropical storm and typhoon advisories and warnings from the 180th meridian westward to the mainland of Asia (as shown in figure 1–9). This responsibility, covering a vast oceanic area, applies to civilian and military interests alike.

Similar to the Fleet Weather Central at Rota, the Fleet Weather Central/JTWC, Guam is another ultramodern, extremely well-equipped, and frightfully busy weather command. In addition to all the standard meteorological and oceanographic equipment and capabilities, the Fleet Weather Central/ JTWC, Guam also has the equipment necessary to receive, record, and transmit weather satellite data and information. It also has an electronic computer laboratory capable of communications of over 4,000 words per minute.

Specifically, this command's mission as assigned by the military Commander in Chief, Pacific, is as follows: (1) provide all tropical cyclone and typhoon advisories and warnings west of 180 degrees longitude, (2) determine typhoon reconnaissance requirements and priorities, (3) conduct investigative and postanalysis programs, including the preparation of annual typhoon summaries, and (4)

conduct tropical cyclone and typhoon forecasting and detection research, as practicable.

The efficiency of any system, whether it be theoretical or operational, is directly proportional to the accuracy, timeliness, density, and frequency of the data fed into the system. Thus, the typhoon-warning system—or any system—can only be as good as the data supplied to the system. The vital aircraft reconnaissance data and information for the typhoon-warning system are obtained by the Navy's squadron AEWRON ONE, flying Lockheed *Warning Star* aircraft and homeported at the Naval Air Station, Agana, Guam and the U.S. Air Force's 54th Weather Reconnaissance Squadron, flying WC-130 (Hercules) aircraft based at Anderson Air Force Base on the northern tip of Guam.

Whenever a tropical depression, a tropical storm, or a typhoon is in existence in the western North Pacific area, serially numbered warnings bearing an *emergency* precedence are broadcast from the Fleet Weather Central/JTWC, Guam every six hours at 0000, 0600, 1200, and 1800 Greenwich time. These warnings include the location of the storm/typhoon, its intensity, and its movement, and also include the predicted intensity, direction and speed of movement, and locations.

In the western North Pacific, the following *Typhoon Conditions of Readiness* apply and are strictly adhered to:

> *Condition FOUR*: Threat of destructive winds possible within 72 hours.
> *Condition THREE*: Destructive winds possible within 48 hours.
> *Condition TWO*: Destructive winds anticipated within 24 hours.
> *Condition ONE*: Destructive winds anticipated within 12 hours.

Civilian and military personnel alike in positions of authority refer to the Fleet Weather Central/ JTWC, Guam as the western Pacific's "guardian angel." No more need be said.

Chapter Two

General Flow of Air about the Planet

The ancient Greeks had a word for almost everything, and the science of weather was no exception. The word for the earth's envelope of air, the *atmosphere*, derives from the Greek words *atmos*, which means vapor, and *sphaira*, which means sphere. To appreciate more fully the intriguing—and sometimes frustrating—large- and small-scale movement of air parcels around our planet, we should know a little about the atmosphere's composition and structure.

COMPOSITION OF THE ATMOSPHERE

Air, the material of which our atmosphere consists, is a mixture of many different gases, as shown in figure 2–1. The chemical composition of dry air is remarkably constant everywhere over the earth's surface and up to a height of approximately 65 miles. A sample of dry air contains about 78 percent (by volume) nitrogen, 21 percent oxygen, almost one percent argon, about 0.03 percent carbon dioxide and minute amounts of other gases listed in figure 2–1.

The air also contains a variable amount of water vapor, most of which is concentrated below 35,000 feet. The maximum amount of water vapor which the air can hold depends entirely upon the tempera-

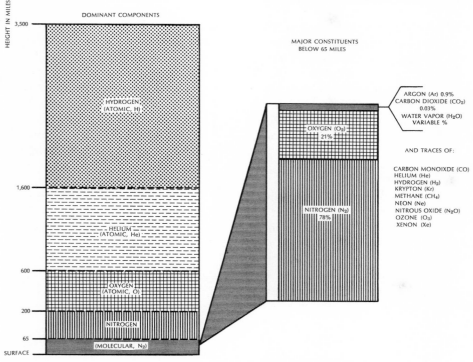

Figure 2–1. Composition of the atmosphere

ture of the air. The higher the temperature, the more water vapor the air can hold. At temperatures below freezing, the amount of water vapor which can be held is very small. At high temperatures, it can amount to as much as four percent (by weight) of the air and water vapor mixture. Water vapor is constantly being added to the atmosphere by evaporation from the earth's surface, particularly from the oceans, lakes, and rivers, and it is constantly being removed from the atmosphere by condensation resulting in precipitation in various forms, but mainly as rain and snow.

Air contains variable amounts of impurities such as dust, salt particles, smoke, other chemicals from sea spray, carbon monoxide, and micro-organisms. Dust, salt, soot particles, and the like are important because of their effect on visibility and especially because of the part they play in the condensation of water vapor resulting in precipitation, in one form or another. If the air were absolutely pure, there would be little condensation. These minute, impure particles act as nuclei for the condensation of water vapor. If the air were perfectly quiet, the heavier particles and gases would settle close to the earth and the lightest would be found the farthest out from the earth's surface. But the constant motion of the air near the surface mixes the gases, so that almost the same proportions exist throughout the atmosphere from the earth's surface upward to about 65 miles.

At a height of about 200 miles, the transition from a predominantly molecular atmosphere to an atomic one occurs, as atomic oxygen (O) displaces molecular nitrogen (N_2) as the dominant component. At roughly 600 miles, atomic helium (He) becomes the major component, and the region beyond 1,600 miles is dominated by atomic hydrogen (H). But in this space era, this is no longer "way out."

STRUCTURE OF THE ATMOSPHERE

All of us, mariner and landlubber alike, live most of our lives at the bottom of a virtual ocean of air which differs in one major way from an ocean of water; water is nearly incompressible. A cubic foot of water on the ocean bottom weighs about the same as a cubic foot of water near the surface. But the air of the atmospheric ocean *is* highly compressible. A cubic foot of air at the earth's surface weighs

billions of times more than a cubic foot of air at the outer edge of the atmosphere. In fact, the atmosphere thins out so rapidly as one leaves the earth that when you are only 3½ miles above the earth, over half the atmosphere, by weight, lies below you. It is primarily in this 3½-mile envelope of heavy air that the weather changes are born.

For convenience, the atmosphere is divided into several layers, or *spheres*, according to their temperature (thermal) characteristics, as shown in figure 2-2. The turbulent layer near the earth in which nearly all weather is embedded, is called the *troposphere*, which varies in thickness from about five miles over the poles to about 11 miles over the equator. In this layer, the temperature generally decreases with height. Above this layer, in the more stable *stratosphere*, the temperature generally increases with height, creating a sort of lid, or cap, which restrains the turbulence of the troposphere. Just above the stratosphere is the *mesosphere*, in which temperature again decreases with height. And finally, the *thermosphere*, the second region in which temperature increases with height, extends upward from the top of the *mesosphere*.

Between each thermal region there is a relatively sharp boundary zone identified by the suffix, *pause*, added to the name of the layer below. Thus, the *tropopause* (at heights of five to eleven miles), the *stratopause* (at a height near 30 miles), and the *mesopause* (at a height near 50 miles) are the upper boundaries of the forenamed layers. This structural description of our atmosphere in terms of its thermal characteristics is just one useful approach. Other terms, found in other weather books, are based on such characteristics as composition, electron density, magnetic control, and so forth. For our purpose, the thermal approach is better.

To the right of the four temperature profile (variation of temperature with height) curves in figure 2-2 is a profile of the speed of sound in air, a value which varies directly as the square root of the kinetic (resulting from molecular collisions) temperature. This variable has a great effect on the lift, drag, and control response of aerospace vehicles, especially at supersonic speeds. For this reason, the *Mach number*, generally defined as the ratio of the speed of the vehicle to the local speed of sound, is widely used to describe vehicle speeds in the supersonic range.

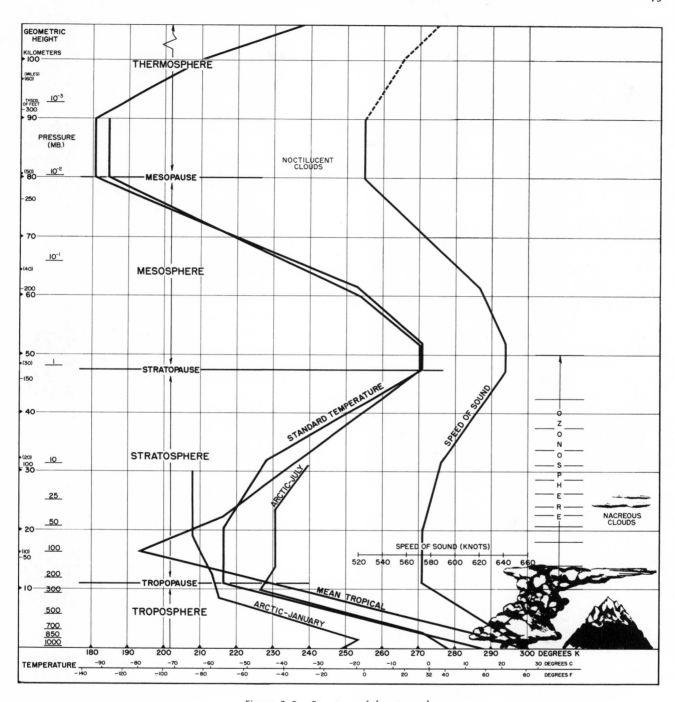

Figure 2–2. Structure of the atmosphere

THE EARTH'S MOTIONS AND SEASONS

As far as we know, the earth has five motions in space: (1) it rotates on its axis once every 24 hours, (2) this rotation has a very slow wobble which takes 26,000 years to complete, (3) it revolves around the sun at 18½ miles per second (making the circuit in 365¼ days), (4) it speeds with the rest of our solar system at 12 miles per second toward the star Vega, and (5) our entire galaxy, with its billions of stars, is rotating in space (our part of it at a speed of 170 miles per second).

Two of these motions directly affect our weather, and their effect is profound. The earth's annual trip around the sun provides the seasons, as shown in figure 2–3, and their typical weather. The earth's daily rotation results not only in day and night; it also produces the major wind belts of the earth, each having its typical pattern of weather.

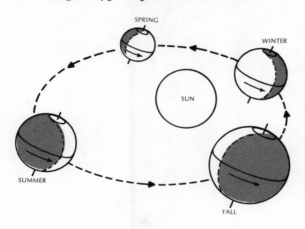

Figure 2–3.   The seasons of the northern hemisphere

The seasons result from the fact that the axis on which the earth spins is slanted 23½ degrees to the plane of its orbit. When the North Pole is tipped toward the sun, the northern hemisphere has summer. The sun's rays beat more directly down on the northern hemisphere, and the days are longer.

The sun is farthest north at the summer solstice, about June 22nd, as shown in figure 2–4. Then the sun is directly over the Tropic of Cancer (23.45 degrees north latitude), and daylight hours are longest in the northern hemisphere and nights are the shortest of any time in the year. The North Pole is in the middle of its annual period of six months of sunlight, and the South Pole is in the middle of its six months of relative darkness.

At the winter solstice, about December 22nd, the North Pole is tipped farthest away from the sun, which is now directly over the Tropic of Capricorn (23.45 degrees south latitude). The southern hem-

isphere has summer, and the northern hemisphere has winter. Conditions are just the reverse of those of the summer solstice; see figure 2–4, again. The Antarctic is now the "land of the midnight sun," and the Arctic is without sun.

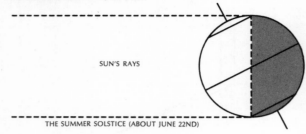

THE SUMMER SOLSTICE (ABOUT JUNE 22ND)

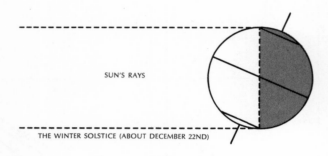

THE WINTER SOLSTICE (ABOUT DECEMBER 22ND)

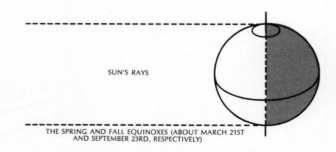

THE SPRING AND FALL EQUINOXES (ABOUT MARCH 21ST AND SEPTEMBER 23RD, RESPECTIVELY)

Figure 2–4.   The solstices and the equinoxes

At the spring and fall equinoxes (about March 21st and September 23rd, respectively), the earth's tilt is sidewise with respect to the sun. Sunlight falls equally on the northern and southern hemispheres. Day and night are of equal duration everywhere on the earth, and the equinoxes mark the beginnings of spring and fall.

Summer is warmer than winter for two reasons: (1) the days are longer, affording more time for the sun to heat the earth, and (2) the sun's rays, striking that part of the earth more directly, are therefore much more concentrated.

Days and nights at the equator are always 12 hours long. The farther north one goes in the summer, the longer the days and the shorter the nights. This is so because the sun in summer shines across the pole, as shown in figure 2–5. The nearer one gets to the pole, the longer the day becomes, until it finally becomes 24 hours long. To an observer in the "land of the midnight sun," the sun appears to move along the horizon, instead of rising and setting.

As winter approaches, the sun seems to move south. The farther north you go, the shorter the winter days and the longer the nights.

But on June 22nd at Washington, D.C., the sun rises farther in the north and at noon reaches about 78 degrees altitude. Its path is much longer and daylight lasts about 15 hours.

The summer sun follows a path more nearly overhead, so that its rays are much more concentrated,

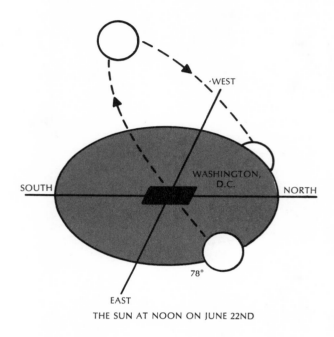

THE SUN AT NOON ON JUNE 22ND

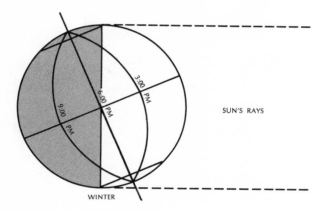

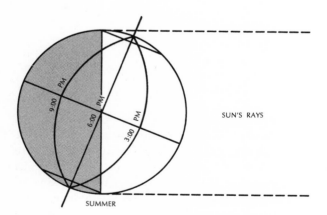

Figure 2–5. The northern hemisphere in winter and summer

The sun's apparent movement relative to the earth in winter and summer is shown in figure 2–6. On December 22nd, at the latitude of Washington, D.C., the sun at noon is only about 26 degrees above the southern horizon. The day is short because the sun's path from eastern to western horizon is short.

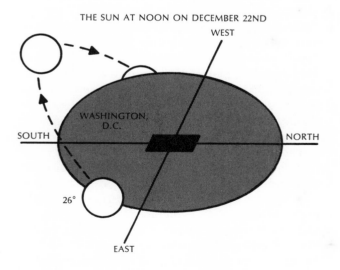

Figure 2–6. The apparent movement of the sun relative to the earth in summer and winter

as shown in figure 2–7. The winter sunlight hits the earth at a much greater slant. The same amount of sunlight now spreads out over a larger area. In addition, winter sunlight must pass through more atmosphere, because it has a more slanted path. More energy is diffused by the atmosphere and less reaches the earth to warm it.

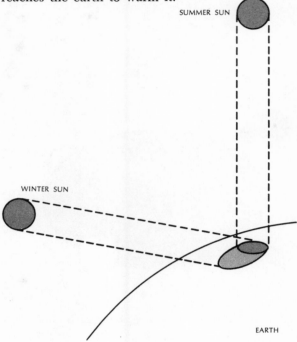

Figure 2–7. Rays of the summer sun and winter sun as they strike the earth

THE EARTH'S ROTATION AND WINDS

The air at the earth's equator receives much more heat than the air at the poles (see figure 2–7, again). Consequently, the warm air at the equator rises—since it is less dense, or lighter—and is replaced by colder—more dense, or heavier—air flowing in from both the north and south sides of the equator. The warm, light air rises and moves poleward high above the earth in both hemispheres. As it cools on moving away from the equator, it sinks, replacing the cool surface air which has moved toward the equator and is being warmed en route.

If the earth did not rotate, the circulation of air over and around the earth would be quite simple, as shown in figure 2–8. Heated surface air would rise over the equator and move north and south high above the earth's surface. En route to the poles, it would cool and sink down at the poles. The sinking air would force the air at the surface toward the equator. So, the surface winds on a smooth,

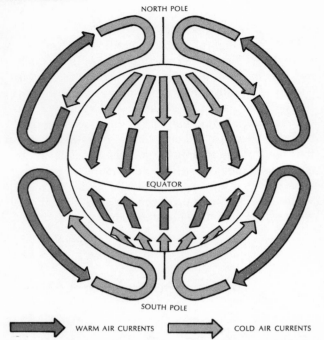

Figure 2–8. Air movement over and around a nonrotating earth

nonrotating earth would all move directly toward the equator with equal speed.

Table 2–1   Approximate Speeds of the Earth's Rotation at Various Latitudes.

Latitude	Knots
90° (Pole)	0
80°	152
70°	295
60°	434
50°	556
40°	669
30°	751
20°	816
10°	855
0° (Equator)	868

The earth's rotation, however, changes this and makes things more complicated. At the equator, the earth is about 21,710 nautical miles around. Because the earth makes one complete rotation every 24 hours, every point on the equator moves eastward at a little more than 868 knots. But every point north or south of the equator moves to the east more slowly, simply because the distance around the earth becomes less as you go north or south of the equator. Exactly halfway between the equator and the poles, at latitude 45 degrees, the earth spins eastward at about 608 knots. At the poles, there is no eastward spin at all, since the poles are the ends of the earth's axis of rotation. These differences in the earth's speed of rotation at various latitudes, as shown in table 2–1, have a marked effect on the winds.

The Horizontal Deflecting Force or Coriolis Force

Considerable deflection of the wind in the northern (and southern) hemisphere is caused by the earth's rotation. A simple experiment will demonstrate this quite clearly.

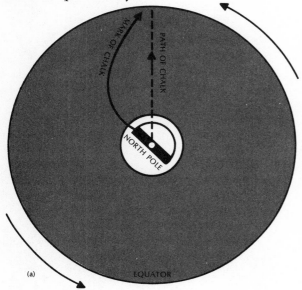

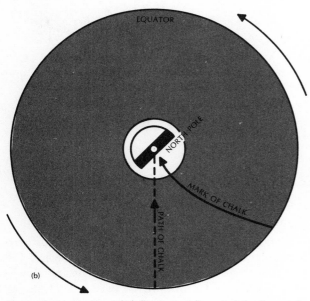

Figure 2–9. Horizontal deflection of the wind (in the northern hemisphere) illustrated by a phonograph record

Not every old salt has a hornpipe, but just about every seagoing man owns a turntable. Place an *old* phonograph record on the turntable. Let the hole at the center of the record represent the North Pole. The rim represents the equator. Now, spin the turntable slowly by hand in a *counterclockwise* direction—the way the earth turns as viewed from above the North Pole. As the record turns, draw a line with a piece of chalk from the center hole (North Pole) directly away from you to the edge of the record (equator). Note how the easterly rotating record causes the chalkline to curve to its right (west as seen from above). In figure 2–9(a), the dotted line shows the actual path of the chalk. The solid line shows the mark your chalk made on the record. Let's repeat the experiment, but draw a straight line from the edge of the record (equator) to the hole in the center (North Pole). Again, as shown in figure 2–9(b), the chalk mark shows a curve to the right.

Winds on the earth curve to the right (in the northern hemisphere) exactly like the chalk lines and for exactly the same reason. In the science of weather, this effect is called the *horizontal deflecting force* or the *Coriolis force*.

Consider a wind blowing from the north and starting at Washington, D.C. Figure 2–10(a) shows this north wind at Washington blowing due south, as viewed by an astronaut in space over the North Pole. The wind blows in a straight line, due south, toward a point P on the earth's equator and also toward a star S out in space. But as the wind moves south, the earth—and point P along with it—is moving eastward. Hence, the straight path of the wind as seen by an astronaut in space becomes a curved path to the right as seen by an observer aboard ship on the earth's surface.

Figure 2–10(a), (b), (c) illustrate the wind curvature which you, as an astronaut, would see from a point in space over the equator. The north wind starts moving due south along meridian X and moves in a straight line directly toward the equator. As the wind moves from Washington (near latitude 39 degrees north) southward to latitude 30 degrees north, the earth's eastward rotation moves meridian X to your right, and meridian Y moves to a position directly under you. By the time the wind has reached the equator, meridian Y has moved east and has been replaced by meridian Z.

As seen from space, the wind still moves directly south as a *north* wind. But to a man aboard ship, or anywhere else on the earth's surface, the wind

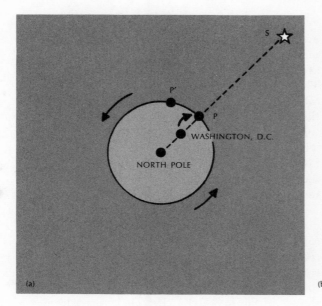

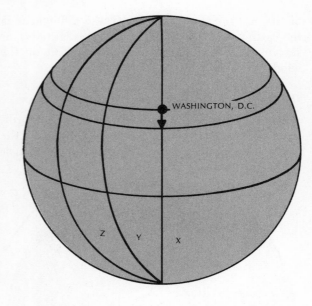

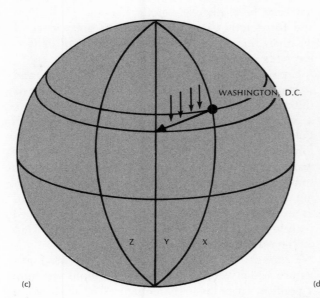

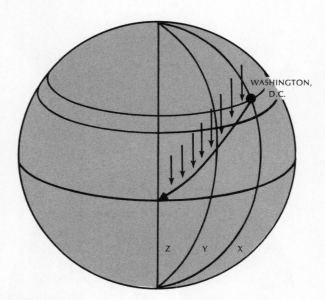

Figure 2–10.   The coriolis force

has taken a curved path and is blowing as a *north-east* wind.

Winds blowing from the south, toward the north, also curve toward their right in the northern hemisphere; this *curving* helps to explain the general circulation of air over the earth.

In the *southern* hemisphere, this effect is exactly reversed, and winds down there curve to the *left*. To see why this is so, repeat the experiment of figure 2–9. But this time, turn the phonograph record *clockwise,* as though you were looking down on the earth from a point in space above the South Pole.

Note how the chalk lines curve to the left.

GENERAL CIRCULATION OF AIR AROUND THE EARTH

The general movement of air in the northern hemisphere begins with air moving toward the north, high above the equator, and curving slowly toward the east because of the earth's rotation (as discussed in the previous section). By the time this upper air has gone about one-third of the way from the equator to the North Pole, it is moving eastward. (See figure 2–11.) As more air from the region of the equator curves north and east into this area (near 30 degrees north latitude), it begins to "pile up," forming an area of high pressure. Some of the air is forced down to the earth's surface, and there, a portion of it flows southward, curving to the right (west) as it travels. This portion forms the so-called *trade winds*, which blow rather steadily from the northeast. But some of the descending air travels northward and is deflected toward the right also, in this case toward the east. This air forms the prevailing westerlies which blow over the middle latitudes (from about 30 degrees to 60 degrees north latitude).

But not all of the high-level air sinks to the earth's surface at 30 degrees north latitude. Some of it continues traveling north, high above the earth. It cools by radiation as it travels and finally contracts and becomes so heavy that it sinks down to the surface near the North Pole. The pressure then builds up in that region, and the cold, heavy air travels southward at the earth's surface, curving toward the right (west) as it travels. At about 60 degrees north lati-

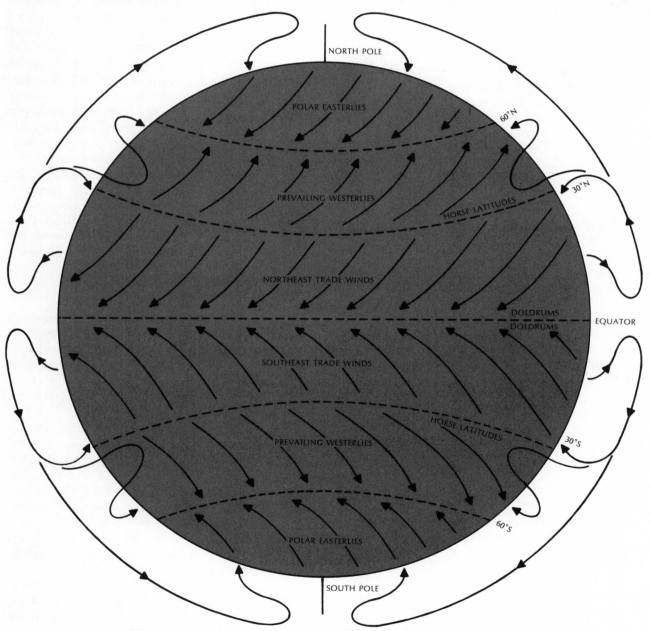

Figure 2–11. Simplified general circulation of air over the rotating earth. Note the reversal of winds in the southern hemisphere. (Compare this with figure 2–8.)

tude, it collides with the prevailing westerlies which are traveling toward the northeast. This zone of collision is called the *polar front*. The warm air moving from the southwest toward the northeast pushes up over the wedge of cold air traveling from the polar regions southward. The rising warm air is cooled rapidly, and unsettled or turbulent conditions result. This polar front is the source of much of the changing, or bad, weather in the United States and over the oceans of the world. The air mass breaks through in cold waves that sometimes reach as far south as Mexico and Florida.

The earth's general circulation, its weather and winds, are modified by many things—the differences in heat absorption of land and oceans, seasonal changes, winds caused by uneven local heating, and so forth. But some general conditions are worth noting. See figure 2–11, again. For example, the circulation in the southern hemisphere is the opposite of the circulation in the northern hemisphere. Down under, the winds curve to the left for the same reason that winds curve to the right in the northern hemisphere. Northern hemisphere wind patterns exist in the southern hemisphere, but they are reversed. Each pattern has its near-counterpart in the southern hemisphere.

Certain basic regions are of particular interest to mariner and landlubber alike. Traveling from north to south in figure 2–11:

The polar easterlies are bounded on the south by the polar front. This zone is one of cold air moving southwest until it collides with the prevailing westerlies. The front, itself, moves as these two air masses push back and forth. At the front, warm, moist air from the south is lifted over the colder and heavier air from the north, and bad weather results.

The prevailing westerlies are the major air flow between about 30 degrees and 60 degrees latitude. These winds are usually from slightly north of west to slightly south of west. If it were not for the repeated invasions of cold air from the areas of the polar front, plus complications caused by air being forced over mountain ranges, weather over the continents between latitudes 30 and 60 degrees would be rather stable because of these winds. Fair weather and clear skies would alternate with steady rains as the air moved northward and cooled slowly. However, the polar front, and other factors which will be discussed later, often cause rapid—and sometimes violent—changes in the weather.

The horse latitudes are a region of relative calm. As the air sinks and is therefore warmed (*adiabat-*

ically—warmed by compression; no transfer of heat), skies tend to be cloudless. Winds are weak and unreliable. It is thought that the term *horse latitudes* originated because sailing ships carrying horses from Spain to the New World often found themselves becalmed with critical shortages of food and water for the animals. As a result, horses were thrown overboard and it is reported that, at times, the sea was littered with the bodies of starved horses.

The trade winds are very well named. These steady northeast (northern hemisphere) and southeast (southern hemisphere) winds marked a very popular route for sailing vessels. Skies are frequently clear near 30 degrees latitude, but cloudiness and heavy, frequent, showery rainfall occur closer to the equator, where the air rises in the doldrum zone.

The equatorial doldrums occur on either side of the equator. As the air at the earth's surface travels toward the equator as northeast and southeast trade winds, it is heated by the direct rays of the sun; consequently, it rises. This rising air creates the equatorial calms, or doldrums. Day after day and week after week may go by without hardly a breeze. Sailing ships try to avoid this region, when possible. From the moist, rising air of this region, the terrifying tropical hurricanes and typhoons are born. This is also the region of extremely heavy tropical rainfall.

WHY THE WEATHER?

Weather is the condition of the earth's atmosphere in terms of heat, pressure, wind, and moisture. These are the basic elements of which weather is made. At great heights where the atmosphere thins out to nothingness, there is no weather. There is no weather on the moon, for example, because it has no atmosphere.

Near the earth's surface, the atmosphere is dense and heavy; here, in the lower atmosphere, we continually see the everchanging, dramatic, and sometimes violent "weather show." But it takes more than just air to make weather. If the earth's atmosphere were never heated, or mixed, or moved around, there would be no weather; more correctly, there would be no weather *changes*. There would be no winds, no storms, no changes in atmospheric pressure, and no rain or snow.

Heat is really the "spoon" that mixes the atmosphere to create weather. All weather changes are caused by temperature changes in different parts of the atmosphere. The sun, which is the source of most of the earth's heat, is a ball of glowing gases some 93 million miles away. This tremendous atomic furnace bombards the earth with something like 126 trillion horsepower of energy every second. Yet this fantastic energy is only about one-half of one-billionth of the sun's total output. Most of this solar energy is lost in space. The sun's energy is transmitted as waves which are similar to radio waves. Some of these are visible light waves; others are invisible. Some waves, although not actually heat waves, change to heat when they are absorbed by objects such as our bodies or the earth's soil. Roughly 43 percent of the radiation reaching the earth strikes the surface and is changed to heat. The rest stays in the atmosphere or is reflected back into space.

Like a greenhouse, the earth's atmosphere admits most of the solar radiation. When this is absorbed by the earth's surface, it is re-radiated as heat waves, most of which is trapped by the water vapor in the atmosphere. Thus, the earth is kept warm. As a thermostat, the atmosphere controls the earth's heat as automatically as any efficient heating system. It protects the earth from too much solar radiation during the day and screens out the dangerous rays. At night, it acts like an insulating blanket which keeps most of the heat from escaping. Without an atmosphere, the earth would experience temperatures similar to the moon's. The moon's surface temperature reaches 212°F during the two-week lunar day. And it drops to about 238°F below zero during the long lunar night. The earth cools faster on clear nights than on cloudy nights, because an overcast sky reflects a large amount of heat back to earth, where it is once again re-absorbed.

Water is always present in the atmosphere. It evaporates from the earth's surface, of which over 70 percent is covered with water. In the atmosphere, water exists in three states—solid, liquid, and invisible vapor. The amount of water vapor in the air is called the *humidity*. The *relative humidity* is the amount of water vapor which the air is holding, expressed as a percentage of the amount which the air *could* hold at that particular temperature. Warm air can hold more water than cold air, and when air with a given amount of water vapor cools, its relative humidity goes up. When the air is warmed, its relative humidity decreases.

Heat evaporates millions of tons of water into the atmosphere daily. Streams, lakes, and oceans send up a steady stream of water vapor. As moist, warm air rises, it slowly cools. Finally, it has been cooled so much that its relative humidity reaches 100 percent. Then clouds form, and under certain conditions, rain or snow begins to fall. This eternal process of evaporation, condensation, and precipitation is called the *water cycle.*

Precipitation such as snow, rain, sleet, and hail can occur only if there are clouds in the sky. But not all kinds of clouds can produce precipitation. Temperature and the presence in the atmosphere of tiny foreign particles (hygroscopic nuclei), or of ice crystals, all help to determine whether precipitation will occur and what form it will take. For instance, snow will not form unless the air is supersaturated (cooled below its saturation point—or dew point, to be discussed later—without its water vapor condensing)—and is considerably below the freezing point of water.

Rain falls from clouds for the same reason that anything falls to earth. The earth's gravity pulls it. But since every cloud is made up of water droplets or ice crystals, why doesn't rain or snow fall from all clouds? The answer is that the water droplets or ice crystals are extremely small. The effect of gravity on them is minute. The air currents move and lift the droplets or crystals so that the net downward movement is close to zero, even though the droplets or crystals are in constant motion. The cloud droplet of average size is only 1/2500 of an inch in diameter. It is so small that it would take over 16 hours to fall a mere 1,000 yards in perfectly still air. It does not fall out of moving air at all. Only when the droplet grows to a diameter of 1/125 of an inch or larger can it fall from a cloud. The average raindrop contains a million times as much water as a tiny cloud droplet. The growth of a cloud droplet to a size large enough to fall out is the cause of rain and the other forms of precipitation. This most important growth process is called *coalescence*—growing together; large and small droplets collide, become bigger, and finally drop as rain.

The layer of the atmosphere next to the earth's surface—the *troposphere*—is the real weather breeder. Here are found nearly all the clouds, and this is where the earth's weather occurs. The air, heated by contact with the earth, rises and is replaced by colder air. These vertical currents that are formed create horizontal winds at or near the earth's surface. Water, evaporated from the land and the seas, rises with the ascending warm air. As the air rises, the surrounding (ambient) pressure lessens, so the air expands. Expansion is a cooling process, and if the air rises high enough, it cools until condensation forms clouds. If the droplets coalesce, precipitation occurs.

Chapter Three

Basic Cloud Formations

Look when the clouds are blowing
And all the winds are free:
In fury of their going
They fall upon the sea.
But though the blast is frantic,
And though the tempest raves,
The deep intense Atlantic
Is still beneath the waves.
      —FREDERIC WILLIAM HENRY MYERS
        (1843–1901)

These few words express so magnificently the wonders and the excitement of the earth's clouds, weather, and oceans. Some of nature's most beautiful artistic displays are the various cloud formations; understanding clouds—how, why, when, and where they form, what shape they take, and what the different types of clouds portend—is important to anyone involved with the weather and its changes. But especially to individuals who follow the sea, a knowledge of clouds is an indispensable tool.

Clouds assume an almost infinite variety of forms. They offer visual evidence of conditions existing in the atmosphere and also of changes which are taking place in the atmosphere. Consequently, clouds also afford an indication of future weather conditions, particularly if they are observed at various times and time intervals to note the changes in cloud structure or type which may be taking place. As one example, one of the first indications of the existence of an approaching midlatitude storm or tropical hurricane or typhoon is the appearance of high-level cirrus clouds. As the storm or hurricane/typhoon moves closer, high-level cirrostratus clouds are observed, followed by lower and thicker altostratus clouds. Finally, with the storm or hurricane at hand, low, dark clouds herald the arrival of high winds and torrential rains.

Most clouds occur almost wholly within the lowest layer of the atmosphere—the troposphere. However, two types of clouds do occur above the troposphere. Clouds may be composed of water droplets, ice crystals, or a mixture of the two, depending on the temperature of the air layer. High clouds are composed mainly of ice crystals. Middle clouds are composed largely of water droplets. Low clouds usually consist entirely of water droplets. Clouds with great vertical development almost always consist of both ice crystals (near the top of cloud) and water droplets (in the lower portion of cloud).

By the use of high-altitude sounding balloons, laboratory experiments, aircraft reconnaissance, meteorological rockets, weather radars, and weather satellites, more and more scientific knowledge is being acquired every day on the fascinating subject of clouds.

CLASSIFICATION OF CLOUDS AND CLOUD NAMES

Fundamentally, clouds are classified according to how they are formed, and there are two basic types. The first is clouds that are formed by rising air currents. These are puffy, or piled-up. They are called *cumulus*, which means accumulated, or piled-up.

The second type is clouds that are formed when a layer of air is cooled below the saturation point without vertical movement. These clouds are in fog-like layers, or sheets. They are called *stratus*, meaning layered or sheet-like.

Clouds are further classified by altitude into five families: (1) clouds above the troposphere, (2) high clouds, (3) middle clouds, (4) low clouds, and (5) vertical clouds. The cloud bases of vertical clouds may be as low as the typical low clouds, but the tops may be at or above 65,000 feet.

The names of clouds are descriptive of their form and type. The prefix *fracto*, meaning fragment, is added to the names of wind-blown clouds that are broken into pieces. The word *nimbus*, meaning rain cloud, is added to the names of clouds which characteristically produce rain or snow. *Alto*, meaning high, is used to indicate middle-layer high clouds of either the cumulus or stratus type. Table 3–1 contains the names of the basic cloud formations, their height ranges, and their associated weather map symbols.

Clouds Above the Troposphere

Noctilucent clouds—the highest clouds of all—are observed at altitudes of 250,000–300,000 feet, far above the troposphere (average height about 45,000 feet) and far beyond the realm of weather. They appear brilliant against the background sky, ultimately becoming a silvery-white. They move at speeds of 100–500 knots, usually from the northeast or east, and are thought to be of fine cosmic dust. They are observed mainly in the polar and subpolar regions.

Nacreous clouds are observed at altitudes of 70,000–100,000 feet, also primarily in polar and subpolar regions. These clouds are of unknown origin and are thought to be composed of tiny water or ice particles.

Table 3–1 Basic Cloud Formations, Height Ranges (in Middle Latitudes), and Weather Map Symbols

(Near the equator, high clouds range from 20,000–60,000 feet, and middle clouds range from 6,500–25,000 feet. Near the poles, high clouds range from 10,000–25,000 feet, and middle clouds range from 6,500–13,000 feet. At all latitudes, low clouds range from near the ground to 6,500 feet.)

Clouds Above the Troposphere	High Clouds	Middle Clouds	Low Clouds	"Vertical" Clouds
300,000 ft. to 250,000 ft.	100,000 ft. to 70,000 ft.	45,000 ft. to 18,000 ft.	18,000 ft. to 6,500 ft.	6,500 ft. to Near the Ground	45,000 ft. to Near the Ground
Noctilucent	Nacreous	Cirrus (Ci)	Altostratus (As)	Nimbostratus (Ns)	Cumulus (Cu)
		Cirrostratus (Cs)	Altocumulus (Ac)	Stratus (St)	Cumulonimbus (Cb)
		Cirrocumulus (Cc)		Stratocumulus (Sc)	

High clouds (18,000—45,000 feet)

DESCRIPTION

Cirrus (*Ci*) clouds (figure 3–1) are detached wisps of hair-like (fibrous) clouds, formed of delicate filaments, patches, or narrow bands. Like the other high clouds, they are composed primarily of ice crystals. They are often arranged in bands which cross the sky like meridian lines and, because of the effect of perspective, converge to a point on the horizon.

MEANING

Cirrus clouds, which are scattered and are not increasing, have little weather meaning except to signify that any bad weather is at a great distance. Cirrus clouds in thick patches mean that showery weather is close by. These clouds are associated with, and formed from, the tops of thunderstorms. Cirrus clouds shaped like hooks or commas indicate that a warm weather front is approaching, and that a continuous-type rain will follow—especially if the cirrus is followed by cirrostratus. They also frequently indicate the presence and location of a jet stream.

Courtesy of R. K. Pilsbury

Figure 3–1.   Cirrus clouds

Courtesy of R. K. Pilsbury

Figure 3–2.   Jet stream cirrus clouds. Note that the condensation trail created by a jet aircraft flying at cloud level disappears in the drier air to the north.

DESCRIPTION

Cirrostratus (*Cs*) clouds (figure 3–3) are transparent, whitish clouds that look like fine veils or torn, wind-blown patches of gauze. They never obscure the sun to the extent that shadows are not cast by objects on the ground. Because they are composed of ice crystals, cirrostratus clouds form large halos—or luminous circles—around the sun and moon.

MEANING

Cirrostratus clouds, when in a continuous sheet and increasing, signify the approach of a warm weather front or an occluded weather front (to be covered in chapter 7) with attendant rain or snow and stormy conditions. If these clouds are not increasing and are not continuous, this means that the storm is passing to the south of you and no bad weather will occur.

Courtesy of R. K. Pilsbury

Figure 3–3.   Cirrostratus clouds with cirrus, cirrocumulus, and lower altocumulus also present

DESCRIPTION

Cirrocumulus (*Cc*) clouds (figure 3–4) are thin, white, grainy and rippled patches, sheets, or layers showing very slight vertical development in the form of turrets and shallow towers. When these clouds are arranged uniformly in ripples, they form what seafaring men call a *mackerel sky.* These clouds are usually too thin to show shadows.

MEANING

Cirrocumulus clouds are quite rare and are of mixed significance. In some areas, these clouds foretell good weather; in others, bad weather. These clouds usually signify good weather along the west coast of the United States, in New England, and in the British Isles. They signify bad weather in most of southern Europe, particularly in Italy.

Courtesy of R. K. Pilsbury

Figure 3–4. Cirrocumulus clouds

Middle Clouds (6,500—18,000 feet)

DESCRIPTION

Altostratus (*As*) clouds (figure 3–5) are greyish or bluish cloud sheets or layers of striated, fibrous, or uniform appearance that cover part, or all, of the sky. These clouds are composed of water droplets, ice crystals, raindrops, or snowflakes. The sun and moon do not form halos, as with the higher purely ice-crystal cirrostratus, but appear as if seen through ground glass.

MEANING

Altostratus clouds are the most reliable weather indicator of all the clouds. They almost always indicate warm air flowing up over colder air in the atmosphere and impending rain or snow of the continuous, all-day type, especially if the overcast cloud layer progresses continually and thickens. These clouds are also a good indication of a new storm development at sea. They frequently signal the formation of a stormy low-pressure area long before it is apparent from sea-level isobars or wind. What is important to seafaring men is that these clouds are a reliable indication of approaching rain or snow, with associated poor visibility, large waves, and heavy swell.

Courtesy of R. K. Pilsbury

Figure 3–5. Altostratus clouds

DESCRIPTION

Altocumulus (*Ac*) clouds (figures 3–6 and 3–7) are most often seen as extensive sheets of regularly arranged cloudlets, white and grey, and somewhat rounded. Through altocumulus clouds, the sun often produces a corona, or disk, usually pale blue or yellow inside, and reddish outside. The corona's color and spread easily distinguish it from the cirrostratus halo, a larger ring covering much more of the sky. Altocumulus clouds are usually composed of water droplets, with ice crystals forming only at very low temperatures.

MEANING

Altocumulus clouds, in general, are significant primarily when they are followed by thicker, high-cloud forms or cumulus-type lower clouds. When arranged in parallel bands, these clouds are found in advance of a warm weather front with associated steady rain or snow. North of this banded-type of altocumulus, altostratus clouds and precipitation should be expected, and if the altostratus is moving southward, rain or snow should be expected. When altocumulus occurs in the form of turrets rising from a common flat base, it is usually the forerunner of heavy showers or thunderstorms.

Courtesy of R. K. Pilsbury

Figure 3–6.   Altocumulus clouds

Courtesy of R. K. Pilsbury

Figure 3–7.   Altocumulus cloud (Lenticularis). This is a classic example of a mountain-wave cloud taken near Taormina, Sicily.

Low Clouds (from near the ground to 6,500 feet)

DESCRIPTION

Nimbostratus (*Ns*) clouds (figure 3–8) are the true rain and/or snow clouds, depending upon the temperature. These clouds are low, amorphous, dark, and usually quite uniform. They are thick enough to blot out the sun, and they have a "wet look." When these clouds precipitate, the rain or snow is usually continuous. These clouds are often accompanied by low *scud* clouds (fractostratus) when the wind is strong.

MEANING

Nimbostratus clouds are of little help as a forecasting tool, since the bad weather is already at hand when these dark clouds with their associated heavy rain or snow are overhead. But if they are at some distance from you and you have a report that they are coming your way, you know what to expect and can take the necessary precautions. Once nimbostratus clouds have formed in your area or are heading your way, bad weather, wind, and sea conditions will persist for many hours. The stormy conditions which almost always accompany these clouds make boating extremely hazardous.

Courtesy of R. K. Pilsbury

Figure 3–8.   Nimbostratus clouds (with fractostratus below)

DESCRIPTION

Stratus (St) clouds (figure 3–9) are low, grey cloud layers, or sheets, with rather uniform bases and tops. These dull, grey clouds give the sky a heavy, leaden appearance. Only a fine drizzle, ice prisms, or snow grains can fall from true stratus clouds, because there is little or no vertical motion in them. Stratus clouds are sometimes formed by the gradual lifting of a fog layer.

MEANING

Stratus clouds do not signify much potential danger. If the wind speed should decrease markedly when stratus clouds are present in large quantity, the base of the cloud could lower to the earth's surface, resulting in a thick fog. Other than light drizzle, no stratus precipitation should be expected except that from higher clouds when stratus forms in advance of a warm weather front. In this case, the rain supersaturates the colder air below the surface of the warm front, and stratus clouds—or perhaps fog—will form.

Courtesy of R. K. Pilsbury

Figure 3–9.   Stratus clouds (with fractostratus present)

DESCRIPTION

Stratocumulus (Sc) clouds (figures 3–10 and 3–11) are grey and whitish irregular layers of clouds with dark patches formed of rounded masses and rolls. These clouds frequently look like altocumulus clouds, but at a much lower level. They are composed of water droplets, except in extremely cold weather. These clouds do not, as a rule, produce anything but light rain or snow.

MEANING

Stratocumulus clouds, which form from degenerating cumulus clouds, are usually followed by clearing at night and fair weather. The roll-type stratocumulus is characteristic of the cold seasons over both land and water, where the air is cooled from below and mixed by winds of 15 knots or more. Stratocumulus will persist for long periods of time under proper air-to-land or air-to-sea temperature relations. Visibility, however, can be seriously reduced in stratocumulus drizzle or snow.

Courtesy of R. K. Pilsbury

Figure 3–10.   Stratocumulus clouds

Courtesy of R. K. Pilsbury

Figure 3–11.   Stratocumulus clouds (with cumulus below)

Vertical Clouds (from near the ground to 45,000 feet)

DESCRIPTION

Cumulus (Cu) clouds (figures 3–12 and 3–13) with little vertical development, are puffy, cauliflower-like clouds whose shapes constantly change. These clouds are brilliant white in the sunlight, often extending from a relatively dark and nearly horizontal base.

MEANING

Cumulus clouds, when detached and with little vertical development, are termed *fair weather cumulus.* (See figure 3–12.) The weather is fine, and nothing hazardous is in the offing. However, when cumulus clouds swell to considerable vertical extent (figure 3–13), heavy showers are likely with associated gusty surface winds in the vicinity of the showers. Since cumulus clouds normally cover only about 25 percent of the sky, they can often be circumnavigated.

Courtesy of R. K. Pilsbury

Figure 3–12.   Fair-weather cumulus clouds

Courtesy of R. K. Pilsbury

Figure 3–13.   Cumulus clouds (with vertical development)

DESCRIPTION

Cumulonimbus (*Cb*) clouds (figures 3–14 and 3–15) are heavy, dense clouds of considerable vertical extent (often to 45,000 feet and higher) in the form of a mountain or huge tower (figure 3–14). These clouds are the familiar *thunderheads*. The upper part of these clouds is usually smooth, sometimes fibrous, with the top flattened to an anvil shape or a vast cirrus plume. These clouds consist of ice crystals in the upper portion and water droplets in the lower portion.

MEANING

Cumulonimbus clouds, or thunderheads, which sometimes reach as high as 65,000 feet, are to be avoided if at all possible. Very gusty surface winds in the vicinity of the thunderstorm, heavy rain, lightning, frequently hail, and in general, a bad time can be expected in the immediate vicinity of these clouds. (See figure 3–14.) If you recognize mammatus development on the underside of the cumulonimbus cloud, as shown in figure 3–15, get out of there as quickly as possible! A tornado or waterspout could possibly develop.

Courtesy of R. K. Pilsbury

Figure 3–14.   Cumulonimbus cloud (with classic anvil)

Courtesy of R. K. Pilsbury

Figure 3–15.   Mammatus development on underside of cumulonimbus cloud

Courtesy of R. K. Pilsbury

Figure 3–16.   Chaotic sky

PRECIPITATION AND OTHER PHENOMENA

Rain was discussed in the final section of chapter 2, "Why the Weather?," and will not be repeated here. Suffice it to say that *any* type of precipitation can occur *only* if there are clouds in the sky, but that not all kinds of clouds can produce precipitation.

Snow is formed when tiny particles in the air act as nuclei upon which water vapor will crystalize. The air must be supersaturated with water vapor and must be below the freezing point (32° F). Microscopic bits of soil, sand, clay, and the products of combustion are common nuclei. In general, cloud temperatures must be from about $+10°F$ to $-4°F$ before snow begins to form. The water vapor changes to snow, however, even without nuclei at high altitudes in supersaturated air at about $-38°$ F.

Snow pellets are sometimes called granular snow and are white in color and of various shapes. Al-

though similar to soft hail, these pellets are too small and too soft to bounce. A single pellet generally forms from many supercooled cloud droplets which freeze together into crystaline form.

Ice pellets consist of transparent or translucent beads of ice, sometimes called *sleet*. This occurs when rain, dropping from high-level warm air, falls through a layer of freezing air. The raindrops first become freezing rain (supercooled), and when striking the ground in this condition, they form glaze. Further cooling produces ice pellets, or true sleet, which bounces upon striking the ground.

Ice prisms form hexagonal plates, columns, and needles that sometimes glitter almost like diamonds as the wind blows them around. Because of their extremely small size, they fall at a very slow rate. Ice needles often make halos around the sun or moon. In very cold climates, ice-needle fogs frequently form at the ground.

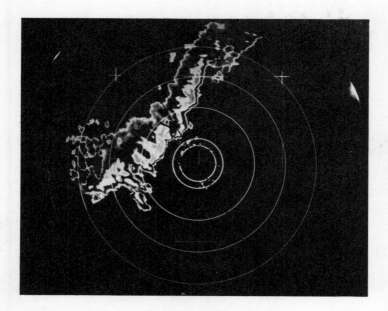

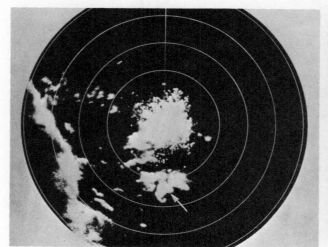 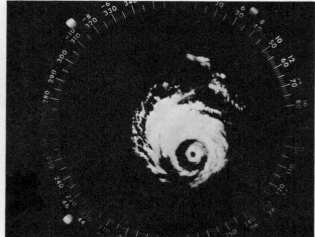

ESSA Photos

Figure 3–17.   Destructive clouds as seen by radar. The distance between radarscope rings is 25 miles. A line of thunderstorms is evident in the top photograph. In the left photograph, the hook-shaped echo (arrow) is a strong indication of tornado formation. The spiral shape of a hurricane is apparent in the photograph at right.

Ice storms are characterized by glaze, and glaze is as destructive as it is beautiful. It occurs when rain or drizzle that has been supercooled (cooled below 32° F but not yet frozen) falls on cold surfaces and immediately freezes. Glaze ice formed from this freezing rain can snap wires and tree branches and cause extremely hazardous—and dangerous—small-boat operations and conditions.

Hail forms as frozen raindrops, high in the clouds, and moves through areas of supercooled water droplets in thunderclouds. Hailstones were thought for many years to develop their onion-like structure by being alternately forced upward by vertical winds in the thunderhead to the freezing level, and then pulled down by gravity to regions where more water was picked up by them. Such up-and-down rides do occur, but the growth of hailstones results mostly from ice pellets picking up water in the supercooled middle and upper portions of the thunderhead. The hailstone layers result from the differences between the freezing rate and the rate at which water accumulates on the pellets.

Frost actually does not fall. Frost is formed when water vapor in the air comes in direct contact with a cold surface and changes directly to small, very fine frost crystals—without condensing into water droplets first (this process is called *sublimation,* as opposed to condensation). Temperatures must be below freezing. Frost crystals grow on the exposed portions of ships, boats, cars, windows, etc., and develop feathery patterns as the primary frost melts and recrystalizes.

Dew does not fall either. Dew forms when water vapor in the air condenses on solid surfaces that have cooled below the condensation point of the air in contact with them. The sweat that forms on boatrails, windows, or the outside of a cool, tall drink on hot days is also dew.

The Barometer and Its Use

Air is one of the physical essentials of life. The condition of the air constitutes a major part of the environment of man, and its diversities are reflected in his activities—man's production increases during periods of high pressure; most suicides occur during periods of low pressure—and culture. And yet, atmospheric pressure (the "push" per unit area), one of the most important of all meteorological measurements used in weather analysis and forecasting, is probably the least detectable by man's senses.

AIR HAS WEIGHT

The atmosphere is held to the earth by the force of gravity. It, therefore, has weight which is manifested by a downward pressure. The weight of a cubic foot of air at sea level is about 1.2 ounces, or .08 pounds. The density—weight divided by volume —of air, then, is 0.08, expressed in pounds per cubic foot. Unlike water, air is highly elastic and compressible. Since the air is fluid and for practical purposes in statical equilibrium, it follows that the pressure at any point in the atmosphere is the same in all directions (downward, upward, and laterally).

A column of air from the earth's surface to the top of the atmosphere exerts a pressure on the earth's surface equivalent to a column of mercury 29.92 inches, or 76 centimeters, or 760 millimeters high, or a column of water 34 feet high.* This pressure is 1,013.25 *millibars* (units normally used on weather maps). This is why mercury, rather than water, barometers are used to measure atmospheric pressure. A 34-foot-tall water barometer would hardly be the instrument for either the seaman or landlubber.

In consequence of the weight of the air, the atmosphere exerts a pressure upon the earth's surface

---

* At latitude 45° and at a temperature of 32°F (0° C) around the measuring instrument.

amounting to about 14.7 pounds per square inch (lbs./sq. in.), or about one ton per square foot. By international agreement, this is defined as *one standard atmosphere*. So we see that air is hardly a light substance. The mass of the atmosphere above a modest-size house with an area of 1,500 square feet is over 1,400 long tons! And a small boat of only 1,600 cubic feet contains over 100 pounds of air.

NORMAL ATMOSPHERIC PRESSURE

Although the atmospheric pressure varies somewhat hour-by-hour and from day to day, the *normal* value is quite well known and can be (and is) expressed in many different units. It is usually given in *pressure* units (units of force divided by area), but it is often given in terms of its balancing column of mercury or water. Table 4–1 is a list of the more common expressions for this normal atmospheric pressure.

Table 4–1   List of Common Expressions for "Normal" Atmospheric Pressure

(*Note: One millibar is 1,000 dynes per square centimeter. A dyne is the unit of force in the centimeter-gram-second system of measurements. The equivalent weight [English units] is 0.0145 pound per square inch.*)

1,013.25	Millibars(mb)
1.01325	Bars (1 Bar = 1,000,000 dynes/cm²)
29.92	Inches of Mercury
760.00	Millimeters of Mercury
76.00	Centimeters of Mercury
14.66	Pounds Per Square Inch
33.9	Feet of Water
1,033.3	Centimeters of Water
1,033.3	Grams Per Square Centimeter
1,013,250.0	Dynes Per Square Centimeter

In middle and high latitudes, barometer readings at sea level usually range between 970 mb. (28.64 in.) and 1,040 mb. (30.71 in.), but readings as high as 1,060 mb. (31.30 in.) and as low as 925 mb. (27.32 in.) may occur occasionally, and substantially lower pressure readings have been known to exist near the centers of tropical hurricanes and typhoons. Table 4–2 consists of a comparison of the various barometer scales.

Table 4–2   A Comparison of Barometer Scales
(* *Normal Atmospheric Pressure*)

Inches	Millibars	Millimeters
31.00	1050.0	787.0
30.50	1032.9	774.7
30.00	1015.9	762.0
29.92*	1013.2*	760.0*
29.50	999.0	749.3
29.00	982.0	736.6
28.50	965.1	723.9
28.00	948.2	711.2
27.50	931.3	698.5
27.00	914.3	685.8
26.50	897.4	673.1
26.00	880.5	660.4

1 Inch = 33.86 Millibars = 25.4 Millimeters

WEIGHT OF THE EARTH'S ATMOSPHERE

Since the normal pressure of the earth's atmosphere is known, it is quite simple to calculate the weight of all the air that surrounds the earth. If we multiply the pressure (in pounds per square inch) by the number of square inches in the earth's surface, the result will be the total weight of the earth's atmosphere. This gives correctly the total *weight* of the atmosphere but does not give the total *mass* quite correctly, because the outer portions have a slightly smaller weight per unit-mass than the inner portions. However, this correction is not very important.

The area of the earth in square inches is
$$A = 4 \times \pi \times r^2$$
$$= 4 \times 3.1416 \times (4{,}000 \times 5{,}280 \times 12)^2$$
$$= 8.1 \times 10^{17} \text{ square inches in the earth's surface}$$
Then, the weight of the earth's atmosphere is
$$W = 14.7 \times 8.1 \times 10^{17}$$
$$= 11.8 \times 10^{18} \text{ pounds}$$
(11,800,000,000,000,000,000 pounds)
$$= 5.9 \times 10^{15} \text{ tons}$$
(5,900,000,000,000,000 tons)

VARIATIONS OF PRESSURE WITH HEIGHT

As mentioned previously, atmospheric pressure is merely the weight of the column of air above a unit-area at the point in question—the column extending to the top of the atmosphere. If a unit-area at a higher elevation is considered, the pressure will be less because there is a smaller quantity (shorter column) of air above it. Thus, the pressure of the air is also a function of the elevation of the unit-area in question. The logical upper limit of the atmosphere is reached when it has become so much rarified that its expansive force and the centrifugal force due to its rotation are equaled by the force of gravity holding it to the earth. The air has no very definite limit, but grows gradually thinner until it becomes imperceptible.

The atmosphere is very compressible, so the lower layers of air are much more dense than the upper layers, and for this reason, the pressure falls very rapidly with an increase in elevation through the lower layers of the atmosphere and much more slowly in the upper layers. So great is the compressibility of the air that one-half of the entire mass of the atmosphere is below 3.5 miles (19,000 ft.), and 97 percent of the mass is below 18 miles (95,000 ft.)! Thus, the top of Mt. McKinley, which has an elevation of 3.8 miles, is above more than half of the earth's atmosphere. When you're jetting about nearly 7.5 miles up (40,000 ft.), considerably *more than half* of the earth's atmosphere is below you!

As a first approximation, we can say that the pressure decreases 1/30th of its value at any given moderate elevation with an increase of 900 feet in height. Starting with a pressure of 30.00 inches (1,015.92 mb.) at sea level, at 900 feet it will have fallen to 29.00 inches (982.05 mb.). During the next 900-foot rise (to an 1,800-foot elevation), it will have fallen 1/30th of 29.00 inches to 28.03 inches (949.20 mb.); the pressure change will continue at about this geometric ratio for each successive change of 900 feet. But the density and weight of the air depend upon its temperature and, to a lesser extent, upon the proportion of water vapor in it, and the force of gravity. Consequently, no *truly accurate* correction for elevation can be made without a consideration of these factors—especially the temperature.

BAROMETERS AND BAROGRAPHS

Pressure measurements from all over the world are essential for the plotting of barometric values on weather maps (along with many other atmospheric measurements) and the analysis and drawing of *isobars*—lines connecting points of equal barometric pressure. The word *isobar* is derived from the Greek, *isos*: equal; and *baros*: weight. By drawing lines connecting points of equal pressure, weathermen determine the pressure patterns all over the world and locate, track, and predict the movements of and changes in, storms, high-pressure areas, weather fronts, low-pressure areas, wind fields, and many other items of great interest to the mariner. Pressure is measured by two types of barometers—*mercurial* and *aneroid*.

Mercurial Barometers

The mercurial barometer is the most accurate of barometers and is used by almost all weather stations around the world. Many larger ships also have a mercurial barometer on board. Mercurial barometers can be of two types, essentially, as shown schematically in figure 4–1. In figure 4–1(a), the glass tube containing the mercury is *U*-shaped, with one end sealed and the other end open to the atmosphere. This is known as the *siphon-type barometer*. The mercury column is supported by the pressure of the air on the open end. When the air pressure is high on the open end of the tube, the mercury column is also high. As the pressure of the atmosphere decreases on the open end, the mercury column lowers. Pressure readings are made either

in inches or in millibars.

The other variation of mercurial barometer, shown schematically in figure 4–1(b), is practically the same—with a few modern refinements—as the one that the Italian physicist Torricelli invented in 1643. Briefly, a long glass tube, a little over a yard long, with one end sealed and the other end open, is filled completely with mercury. The open end is then temporarily sealed (with a cork or finger), inverted, and placed in a cistern partly filled with mercury. When the cork or finger is removed, the mercury in the tube will sink somewhat and come to rest at a level of about 30.00 inches above the level of the mercury in the cistern. There is, then, a vacuum above the mercury in the tube and, therefore, no atmospheric pressure above the mercury within the tube. Since the atmospheric pressure acts on the free surface of the mercury in the cistern, it is clear that the weight of the mercury column above the free surface of the mercury in the cistern must be equal to the weight of the air column above the same surface. The length of the mercury column at any moment indicates the atmospheric pressure, and it is measured by means of a scale placed alongside the glass tube. This is called the *cistern-type barometer*. Weather station mercurial barometers are precision-made instruments and are accurate to 1/1000 inch. Figure 4–2 is a photograph of one of the latest types of mercurial barometer.

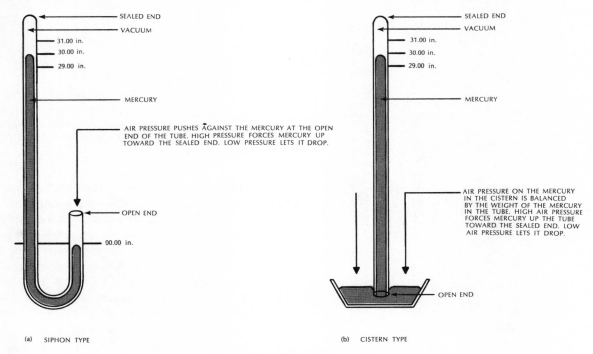

Figure 4–1. Mercurial barometers

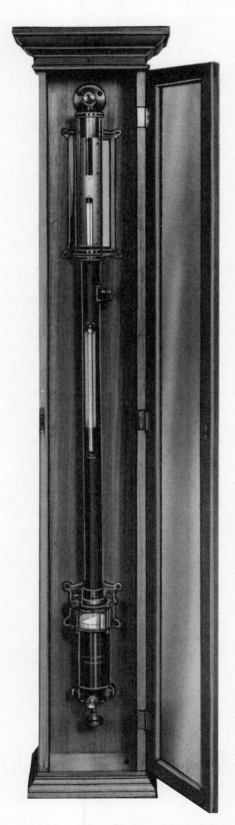

Courtesy of Science Associates
Figure 4–2. Precision observation barometer

Aneroid Barometers

Basically, an aneroid barometer (a barometer without fluid), shown schematically in figure 4–3, is a corrugated metal container from which the air has been removed (a vacuum exists). The corrugations and a spring inside the container prevent the air pressure from collapsing the container, or chamber, completely. As the air pressure increases, the top of the container bends in. As the air pressure decreases, the top of the container bows out. Gears and levers then transmit these changes to a pointer on a dial.

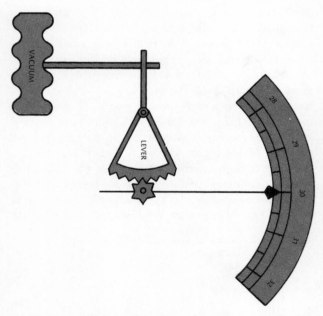

Figure 4–3. Aneroid barometer. Increasing air pressure pushes down the top of the vacuum chamber. This pulls down the attached lever and works the gears, which move the pointer to the right. Decreasing air pressure causes the pointer to move to the left.

Although aneroid barometers are not as accurate as mercurial barometers, they are much cheaper and much more robust since atmospheric pressure is balanced by elasticity forces rather than by mercury in a long glass tube. Consequently, they are ideal pressure instruments for use on board smaller sea-going vessels and boats. Because of the relation between pressure and altitude (described previously), aneroid barometers are also used as altimeters. In this case, the dial is marked in feet to show *altitude* instead of *pressure*. Figure 4–4 is a photograph of one of the newest models of aneroid barometers.

Courtesy of Science Associates
Figure 4–4. Navy type barometer

The Barograph

A barograph is simply an aneroid barometer which has been configured so that atmospheric pressure changes are *recorded* on a paper-covered rotating drum. As shown in figure 4–5, a date/time-pressure record sheet is wrapped around a cylinder or drum. Inside the cylinder is a clock mechanism which rotates the cylinder once every seven days (or some other number of days). The pen arm is actuated by atmospheric pressure changes (up and down) as the cylinder rotates, and it writes a continuous record of pressure on the graduated sheet of paper.

Courtesy of Science Associates
Figure 4–5. Utility barograph

BAROMETER CORRECTIONS

For barometer readings to be made comparable around the world and to be significant, the effects of the variation of gravity between different latitudes, of height above sea level, and of temperature of the instrument must be eliminated. This is done by making small corrections to barometer readings, and the process is known as the *reduction of barometric pressure to standard conditions.*

Temperature Corrections

Mercury expands as it increases in temperature, so that the same atmospheric pressure is balanced by a longer column of mercury on a warm day and by a shorter column on a cold day. A correction is applied to obtain the reading of the barometer that it would show at the standard temperature. Depending on the particular type of barometer used, this standard is either 12°C (54°F) or 0°C (32°F).

Altitude Corrections

Atmospheric pressure, which is a measure of the weight of air above any point, will naturally be less on the bridge of a ship than at sea level. In order to make pressure readings all comparable around the world, the pressure that is read on the instrument needs to be corrected to obtain the pressure that the instrument would record assuming it could be located at sea level in the same place. The actual difference depends not only upon the height of the instrument, but also upon the density of the air between it and sea level. The density of the air in the column between the instrument and sea level varies according to the temperature of the air in that column, and this is obtained by reading the temperature of the air (dry-bulb).

Latitude Corrections

Because of the flattening of the earth at the poles and the fact that the vertical component of the centrifugal force due to the earth's rotation (which acts in the opposite direction from gravity) is greatest at the equator, the force of gravity increases steadily in going away from the equator toward the poles. The standard value of gravity is taken as 980.665 centimeters/second^2, which is practically the same as that at sea level in a latitude of 45 degrees. Thus, a *negative* correction must be applied in low latitudes, where the barometer always reads "high" by a small amount because the "weight" of the mercury is less than at latitude 45 degrees. A small *positive* correction must be applied in latitudes higher than 45 degrees.

Index Corrections

Even after all the above corrections have been made, it is found that the readings from individual barometers still differ slightly because of the differing values of capillarity of the mercury. *Capillarity* is the action by which the surface of a liquid, where it is in contact with a solid, is elevated or depressed, depending upon the relative attraction of the molecules of the liquid for each other and for those of the solid. These index corrections are determined individually for every barometer, the amount of which is clearly stated on the barometer's certificate.

Barometer Correction Example

Assume that you are aboard ship, on the bridge, at latitude 27 degrees. The height of the bridge is 53 feet, the thermometer attached to the barometer reads 77°F, the true temperature of the air is 78°F, the index correction of the barometer is +0.3 millibars, and your mercurial barometer reads 1017.3 millibars (30.041 inches). Standard conditions are 32°F (0°C); gravity of 980.665 cm/sec.² What is the true sea-level pressure reading? Refer to either the *Smithsonian Meteorological Tables,* the *Marine Observer's Handbook,* or another suitable publication.

	Millibars
Uncorrected Barometer Reading	1017.3
Index Correction	+ 0.3
	1017.6
Temperature Correction for 77°F	− 4.3
	1013.3
Height Correction for 53 ft.; air temp. 78° F	+ 1.8
	1015.1
Gravity Correction in Lat. 27 degrees	− 1.6
CORRECTED BAROMETER READING	1013.5

Aneroid barometers need only to be corrected for index error and for altitude. These instruments work on the principle of the balancing of atmospheric pressure by a force due to the elastic deformation of a strong spring attached to a metal box from which the air has been exhausted and which is thereby prevented from collapsing under the external pressure of the atmosphere. Therefore, aneroid barometers do not need to be corrected for changes in gravity. They normally include a device for compensating for small changes of temperature which is ensured either by using a bi-metallic link or leaving a calculated small amount of air in the vacuum chamber. Aneroid barometers that are compensated for temperature are usually so marked.

The readings of aneroid barometers after correction, as indicated immediately above, should be compared as frequently as possible with corrected readings of mercurial barometers. The reason for this is that the index error of all aneroid barometers is likely to change quite frequently due to changes in the elasticity of the metal of the vacuum chamber. Any U.S. Weather Bureau office will be glad to make barometer comparisons for you at no cost. Foreign weather offices will also render this service.

DIURNAL VARIATION OF PRESSURE

The frequently occurring changes of pressure as shown by a barograph trace may be due to many causes, and in middle latitudes in winter, it is not possible to discern any systematic variation. But in lower latitudes, and in the summer months in middle and higher latitudes, a diurnal variation of pressure with a definite pattern becomes evident. Over a long period of time, the mean daily pressure range is about one millibar in middle and higher latitudes. But in lower latitudes and in the tropics, sometimes two to three millibars are observed. Figure 4–6 illustrates a typical diurnal pressure curve in the lower latitudes, with maximum values at 10:00 a.m. and 10:00 p.m. (local time) and minimum values at 4:00 a.m. and 4:00 p.m. (local time). These result from the effects of atmospheric tides.

In middle and higher latitudes, the irregular pressure changes are usually so much larger than the diurnal variations that the latter need not be taken into account, as a rule. In lower latitudes and in the tropics, these irregular changes are usually much smaller than the diurnal change from which they can hardly be distinguished, until the known diurnal change has been subtracted. This procedure is most important in tropical regions, because a fall of pressure of two or three millibars below the known value of the diurnal change of pressure may be the first indication of the generation or the approach of a tropical hurricane or typhoon. One should also be on the lookout for cirrus and cirrostratus clouds and an above-normal, long-period swell.

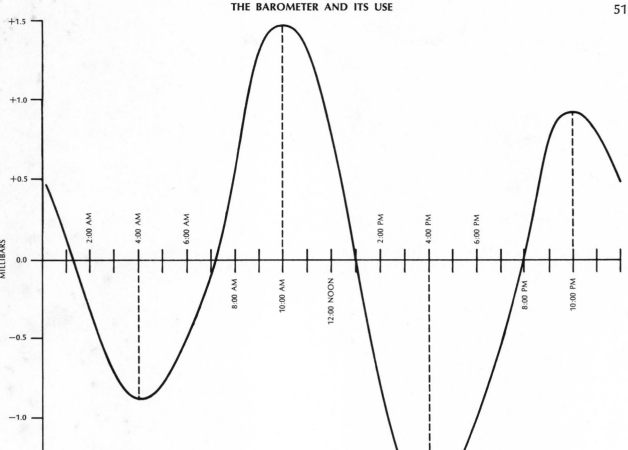

Figure 4-6. Diurnal variation of pressure in lower latitudes

PRESSURE TENDENCY

The name *tendency* is given to describe the *rate of change* of pressure with time. In standard meteorological practice, it usually refers to the time interval of three hours before the pressure observation is made. It is most important for weathermen and mariners to know the change and rate of change of pressure. A ship's report of barometric tendency, of course, does not tell this pressure change directly, because part of the tendency is due to the ship's progressive movement. If the ship is at anchor or in port, the tendency is representative. However, by adjusting for the ship's direction and speed of movement, the true barometric tendency can be easily calculated, if one has the latest weather map available.

Corrected readings of barometric tendency help weathermen and mariners to determine the movements and intensity changes of the various pressure systems, since the barometer usually falls in advance of a low-pressure system (and to the rear of a high-pressure system) and rises in advance of a high-pressure system (and to the rear of a low-pressure system). If the pressure system is stationary, the tendency indicates whether the pressure system is increasing or decreasing in intensity. This will be discussed in detail in a later chapter.

Lines drawn on a weather map through points of equal barometric tendency are called *isallobars*. These isallobars are of great value to weather forecasters because they show the rate of change of pressure with respect to time all over the globe.

Chapter Five

Significance of High-Pressure and Low-Pressure Areas

As discussed in chapter 2, the unequal heating of the earth between the equator and the poles causes meridianal (north-south) winds. The rotation of the earth turns the winds to the *right* in the *northern* hemisphere and to the *left* in the *southern* hemisphere. The result is that this movement of great masses of air creates the overall pattern of the earth's air circulation. But it does something else, too. It creates huge whirling masses of air called high-pressure cells, or simply *highs,* or anticyclones; it also creates low-pressure cells, or simply *lows,* depressions, or cyclones.

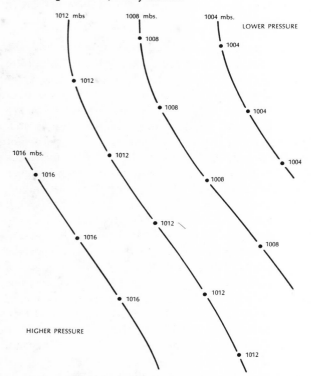

Figure 5–1.   Isobars on a surface weather map

ISOBARS

Atmospheric pressure readings are taken simultaneously all over the world, and then these readings are *reduced to standard conditions* as described in chapter 4. When these values are plotted on charts, lines are drawn connecting points of equal pressure. (See figure 5–1.) These lines of equal (or constant) pressure, called *isobars,* resemble equal-altitude contours which define the hills and valleys on a topographic chart, as shown in figure 5–2. A chart for a particular time which includes such data as wind, weather, temperature, clouds, isobars, etc., for a large number of stations is called a weather map, or more frequently a *synoptic chart,* because it gives a synopsis or general view of the weather conditions over a large area (or the entire globe) at a given instant of time. Isobars are usually drawn at intervals of four millibars (mbs.), i.e., 1012, 1016, 1020 mbs., etc., on weather maps. Any synoptic chart will show a distribution of pressure in which there are regions of high pressure and low pressure resembling the mountains, hills, ridges, and valleys found on topographic charts. A typical region of high or low pressure on a weather map has a rather compact shape, may be 1,000 to 2,000 miles in diameter or width, and is seldom less than a few hundred miles in width.

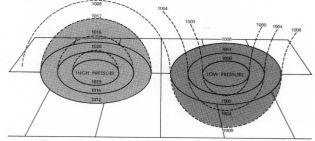

Figure 5–2.   A three-dimensional view of isobars

PRESSURE GRADIENT

The idea of *pressure gradient* is probably best illustrated by reference to a topographic countour map, where at any particular place or point we can estimate the gradient, or slope, of the ground. The steepest slope is seen to be at a right angle to the contour lines, and its numerical value is greater where the contour lines are close together. Similarly, we speak of the *pressure gradient* as being at a right angle to the isobars, its magnitude being measured by the ratio: pressure difference/distance, in suitable units. Thus, if two adjacent isobars (at a four-millibar interval) are 50 miles apart, the pressure gradient is 4/50, or 0.08 mb./mile between them, at a right angle to the isobars, and in the direction *from high toward low* pressure.

PRESSURE AND WIND

Fundamentally, the wind (simply air in motion) is a balance of three forces (if we exclude curvature). Referring to figure 5–3, these forces are the coriolis (horizontal deflecting) force *C*, the frictional force *F*, and the pressure force *P*. Arbitrarily assume that a west wind is blowing (*W*). The coriolis force *C*, we know from chapter 2, is at a right angle to the wind (and pulling to the right in the northern hemisphere). The frictional force *F* is

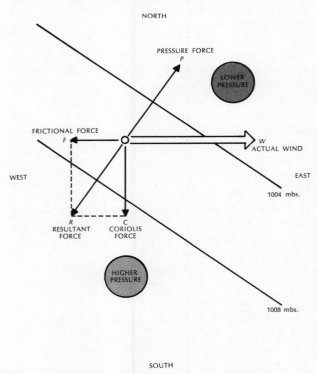

Figure 5–3. The actual wind as a balance of three separate forces

opposite to the direction of the wind *W*. The resultant (combined effect) of the coriolis force and the frictional force is indicated by the arrow *R*. So, to obtain balanced motion, the pressure force *P* must be equal to the resultant *R*, but must have a direction opposite to that of *R*. Since the pressure force is perpendicular to the isobars (acting in the direction from higher toward lower pressure), we see from figure 5–3 that the wind *W* at the earth's surface must blow at an angle across the isobars, in a direction from higher toward lower pressure. This is most important to remember.

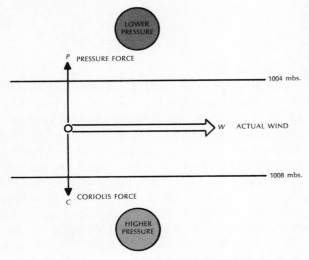

Figure 5–4. Direction of the actual wind above the surface frictional layer

The angle between the wind direction and the isobar increases with increasing friction. Therefore, the angle is greater over land than over water. Over land, this angle averages about 30 degrees; over water, about 15 degrees. As we ascend from the earth's surface, the frictional force diminishes and, for all practical purposes, is negligible at a height of 3,000 feet. At this level, then, the coriolis force is balanced by the pressure force, as shown in figure 5–4, and the wind blows parallel to the isobars. Thus, the surface isobars are indicative of the wind direction and speed at the 3,000-foot level. Since the frictional force decreases with increasing height, the wind velocity increases as we ascend through the atmosphere. Figure 5–5 illustrates the variation of wind speed and direction with height. Observations have shown that the surface wind speed over land is about 40 percent of the wind speed at 3,000 feet, whereas over the oceans it is about 70 percent. The surface wind over land crosses the isobars (from higher toward lower pressure) at an average angle of 30 degrees; over the water it is 15 degrees. The

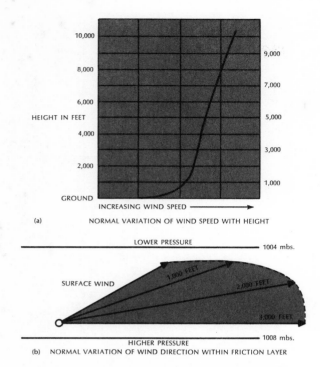

(a)   NORMAL VARIATION OF WIND SPEED WITH HEIGHT

(b)   NORMAL VARIATION OF WIND DIRECTION WITHIN FRICTION LAYER

Figure 5–5. Normal variation of wind speed and direction with height

wind at 3,000 feet blows parallel to the isobars. (See figure 5–6.)

From what we have discussed thus far, it is obvious that if one stands with his back to the wind (in the northern hemisphere), low pressure will always be to his left, and high pressure to his right. In the southern hemisphere, the situation is reversed. This law was first formulated in 1857 by the Dutch meteorologist Buys Ballot and bears his name. This law is also known as the *Baric Wind Law*.

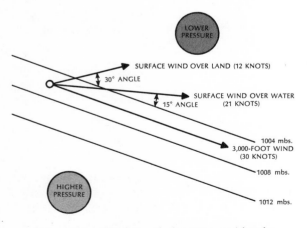

Figure 5–6. Wind speed, direction, and height

PRESSURE SYSTEMS

If we look at a weather map or a series of weather maps and study the isobars over a large area, we notice almost at once that there is only a limited number of types of pressure patterns (or systems) that occur in nature. Figure 5–7 shows diagrammatically the various types of pressure patterns most commonly observed on weather maps and the corresponding winds in the northern hemisphere.

The principal types of pressure patterns (or systems) are the *highs* (or high-pressure cells, or anticyclones) and the *lows* (or low-pressure cells, cyclones, or depressions). As one would expect, a *low* is defined as an area within which the pressure is low, relative to the surroundings. From figure 5–7, we see that the wind circulation around a low is counterclockwise (in the northern hemisphere), with the wind crossing the isobars at an angle, blowing toward lower pressure.

A *high* is defined simply as an area within which the pressure is high, relative to the surroundings. The wind circulation is clockwise around an anticyclone (in the northern hemisphere), with the wind crossing the isobars at an angle and blowing from higher toward lower pressure.

A *trough* of low pressure is an elongated area of relatively low pressure which extends from the center of a cyclone. The trough may have U-shaped or V-shaped isobars (see figure 5–7). The V-shaped isobars are associated with weather fronts, which will be discussed in chapter 7. The wind circulation around a trough is essentially cyclonic (counterclockwise).

A *ridge* (or wedge) of high pressure is an elongated area of high pressure that extends from the center of an anticyclone (high). The wind circulation is essentially anticyclonic (clockwise).

A *col* is the saddle-backed region between two anticyclones and two cyclones, arranged as shown in figure 5–7.

In all the surface pressure systems, the winds blow at varying angles across the isobars, as described in the previous section, in the direction from higher toward lower pressure. Thus, the winds blow clockwise and spiral outward from the centers of anticyclones (in the northern hemisphere), while those of a cyclone blow counterclockwise and spiral in toward the center. In the very center of pressure systems, the pressure gradient vanishes, and there are either calm conditions or light, variable winds. The same is true of a col.

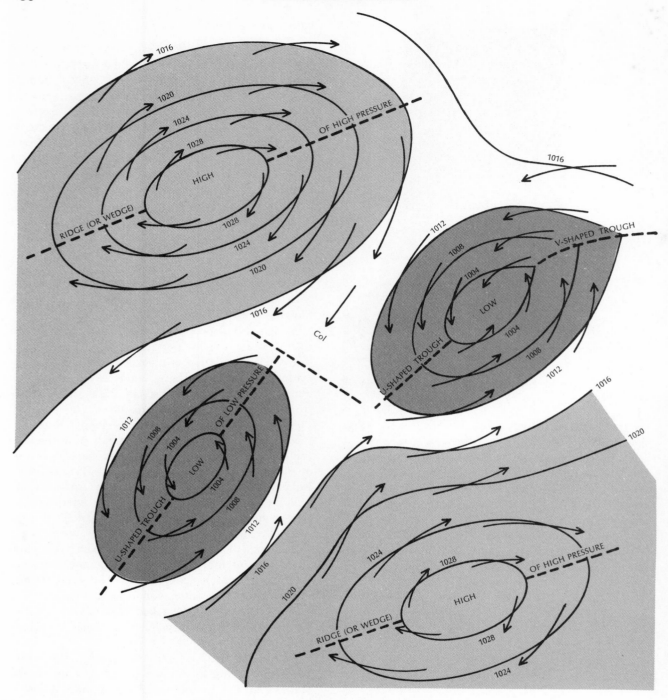

Figure 5–7.   Various types of pressure systems

ANTICYCLONES

There are three basic theories regarding the formation of anticyclones (or highs), but for our purpose, arguments and counterarguments regarding the validity of these theories are not very important. Suffice it to say that local high-pressure areas may develop where air is cooled, compressed, and caused to sink from higher altitudes to the earth's surface

(called *subsidence*). Figure 5–8(a), (b) shows some of the more important features of anticyclones in plan view and vertical cross-section respectively. It must be realized that since surface barometric pressure varies by only a few percent (rarely more than three percent in 24 hours; pressure changes in the vertical are *much* larger), air converging in one layer of the atmosphere must be nearly *compen-*

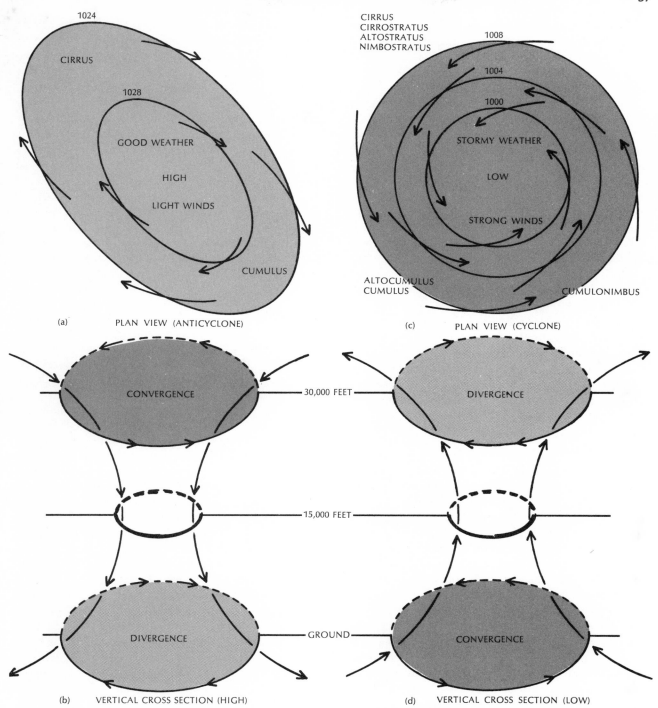

Figure 5–8. Components of cyclones and anticyclones

sated by air diverging in another layer. Otherwise, surface pressure would rise or fall almost without limit.

Because the *mass* of the air is conserved, the air must move vertically from the layers of convergence to the layers of divergence. Figure 5–8(b), (d) illustrates schematically this convergent and diver-

gent motion, and the vertical currents in highs and lows. When air converges aloft as in figure 5–8(b), it moves downward (subsides), bringing good weather with it, because the air is compressed and heated adiabatically and the relative humidity decreases markedly. When it diverges aloft as in figure 5–8(d), the upward movement of the air leads to the formation of clouds and precipitation. Since the airflow becomes increasingly counterclockwise in a convergent layer and clockwise in a divergent layer, cyclonic and anticyclonic flow must alternate with height.

Anticyclones show as much variation as people do. They range in size from the tiny ones only 200 miles across to the huge ones over 2,000 miles in length. And a "granddaddy" every now and then will cover almost the entire United States. Some highs are nearly circular in shape; others are elliptical. But near the center, almost all highs are characterized by clear skies, light wind, and good weather.

Although the speed of movement of anticyclones varies widely (and sometimes they remain stationary for days), during the summer months a good average movement would be 390 nautical miles per day (16 knots). During the winter months, the average movement of highs is somewhat greater, about 565 nautical miles per day (23.5 knots). Surface wind speeds increase considerably, going outward from the center toward the periphery of an anticyclone (i.e., the pressure gradient increases).

Anticyclones are usually classified as either *cold* or *warm* by weathermen.

Cold Anticyclones

A cold anticyclone is one in which the air at the surface and in the lower layers of the troposphere is colder than the air surrounding the high. The air in the anticyclone is thus more dense than the surrounding air, level for level. The high pressure of a cold anticyclone is therefore due primarily to the density of the lower layers of the troposphere being greater than the density of the same layers in the area surrounding the anticyclone.

Examples are the "semi-permanent" anticyclones over the continents in winter. These anticyclones occur most often over Siberia and North America. They are not strictly permanent because these areas are occasionally invaded by traveling low-pressure areas (depressions). But after each period of cyclonic activity, high pressure tends to be re-established, and may then exist for weeks with little change.

These seasonal anticyclones control the atmospheric circulation over wide areas. They are the source regions of continental polar air masses (to be covered in the next chapter). Countries that normally enjoy mild maritime conditions occasionally experience the rigors of continental winter dur-

ing an invasion of continental polar or arctic air. In the British Isles, this sometimes occurs in winter months when a separate anticyclone develops over northern Europe, resulting in persistent easterly winds across the North Sea and the British Isles.

On the east coasts of North America and Asia, the outflow of continental polar air maintains a sharp temperature contrast off the coast, where a warm ocean current runs. These temperature contrasts result in the formation and development of low-pressure cells, which then travel eastward.

Although cold anticyclones are of limited vertical extent—seldom exceeding 10,000 feet—they play a very important role in the lower level atmospheric circulation in winter. In addition to these seasonal cold anticyclones, other more transitory ones also exist.

Warm Anticyclones

In a warm anticyclone, the air throughout the greater part of the troposphere is warmer, level for level, than its environment. Near the surface, the air in a warm anticyclone therefore differs little in density from that of the surrounding air. Since surface pressure is dependent on the total mass of the air above the surface where it is measured, the excess surface pressure in a warm anticyclone must result from air of greater density than the surrounding air being present at higher levels in the atmosphere. This is brought about by the convergence of air at the higher levels, accompanied by subsidence in the lower levels. Also, in warm anticyclones, the pressure at higher levels is greater than in the surrounding air, level for level, since pressure falls more slowly with height through a column of warm air than through a column of cold air.

Examples of warm anticyclones are the oceanic subtropical belts of high pressure (the *horse latitudes* at 30 degrees latitude). They are an essential feature of the general circulation and cover the areas where the air is subsiding. Consequently, the weather there is generally fine, with few clouds and good visibility. The subtropical anticyclones are the source regions of tropical maritime air masses, providing the warm air which feeds the traveling lows of the middle latitudes. The warm anticyclones tend to migrate slightly northward in summer and slightly southward in winter.

Upper-air observations made above anticyclones show that the air is relatively dry, which is to be expected in view of the subsidence taking place. They may also show an inversion of temperature (increase of temperature with height) known as a *subsidence inversion*. As a result, the highest temperature in an anticyclone is often found at a height of 1,000 to 2,000 feet above the ground. This is especially true over the sea. To some extent, this inversion aids fog formation. At sea, this happens because the moisture evaporated from the sea is kept in the lowest layers, particularly if the air stream is moving toward higher latitudes; that is, toward progressively colder sea temperatures. Over land areas, other factors, such as radiation to clear skies and high moisture content, are assisted by the absence of wind associated with the inversion to produce a drop in temperature sufficient to cause fog by condensation.

Clouds forming in the lower layers of an anticyclone, as for example, when moist air from the sea flows over land that is being heated by the sun, tend to spread out at the temperature inversion, forming a layer of stratocumulus clouds. This is more typical of the periphery of an anticyclone than of the center, where the subsidence usually prevents any cloud formation.

CYCLONES

Extra-tropical cyclones (lows of non-tropical origin) frequently form along weather fronts. Hence, they occur with the greatest frequency in the higher mid-latitudes where the cold air masses and warm air masses meet along the polar and the arctic weather fronts. In the northern hemisphere, there is a maximum frequency of lows near 50 degrees north in winter, and near 60 degrees north in summer. In the Atlantic, one of the most favored regions for the development of lows is off the Virginia coast and in the general area to the east of the southern Appalachians. These lows sometimes undergo almost explosive intensification and are often called "Hatteras Storms." They move northeasterly along the Gulf Stream. Sometimes they eventually stagnate near Iceland or in the waters between Greenland and Labrador. Other times they produce dangerous storm conditions around the British Isles after having crossed the North Atlantic.

In the Pacific, there is a broad band of frequent cyclone activity extending all the way from Southeast Asia to the Gulf of Alaska. During the winter months, these storms become very intense, and usually travel northeastward to accumulate in the Gulf of Alaska. Some storms, which form on the mid-Pacific polar front, take a more southerly track and eventually reach the coast as far south as Southern California.

Figure 5–8 (c), (d) shows some of the more important features of cyclones in plan view and vertical cross-section, respectively.

Like anticyclones, cyclones exhibit a large variation in both size and shape. The smaller lows may be only a few hundred miles across, whereas the larger ones may extend to 2,000 miles or more. There have been occasions when the entire North Atlantic weather map was dominated by a single, violent low-pressure cell reaching all the way from the British Isles to Newfoundland! Some lows are very nearly circular in shape, while others are strongly elliptical.

Like highs, lows travel at varying speeds, and they too sometimes remain stationary for a day or two. During the summer months, lows move with an average speed of about 18 knots (432 nautical miles per day). During the winter months, they travel somewhat faster at an average speed of about 25 knots (600 nautical miles per day). Lows are much more stormy and much more sharply defined in winter than in summer.

Low-pressure systems are characterized by many types of clouds, ranging from high clouds to low clouds, moderate to heavy precipitation in the form of rain or snow (depending upon the temperature), strong winds that sometimes shift abruptly, high seas, and generally stormy conditions.

Since both highs and lows travel generally in an eastward direction, in the zone of prevailing westerlies, any high or low to your west will probably move over your position. This is where a certain amount of treachery enters the picture. Highs and lows *may* remain almost stationary for a period of time; they *may* fade out; they *may* move north or south—or even retrograde to the west. So the general rule of eastward movement has only limited usefulness. After we have mastered the facts about air masses (chapter 6) and weather fronts (chapter 7) and learned to read weather maps in a subsequent chapter, we shall be able to use this general rule much more prudently.

Formation of Lows

As shown in figures 5–9 through 5–12, lows are usually formed by one of four methods. Referring to figure 5–9, the major lows are formed by horizontal wave-like action where two different air masses come together (between two highs of different temperature), coupled with rising air currents (vertical motion). The "wave" grows larger and larger and finally breaks like an ocean wave.

The whirling air creates a low-pressure cell, and frequently a storm is born. This complicated process by which the major lows are formed is discussed in some detail in chapter 7.

A local low may form when the air under a cumulonimbus cloud (thunderhead) is rising very rapidly, as depicted in figure 5–10. This low-pressure area is generated as the air surrounding the region directly beneath the cumulonimbus cloud spirals horizontally inward in a counterclockwise fashion because of the earth's rotation, and then rises very rapidly to great heights in the atmosphere. Such lows are quite small in diameter, averaging only 20 to 35 miles across.

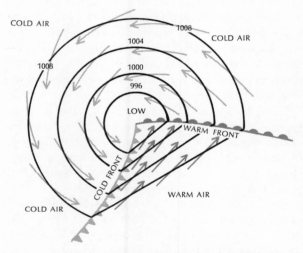

Figure 5–9. Low formed by horizontal wave-like action between two air masses of different temperature

A heat low, as shown schematically in figure 5–11, develops over deserts and other intensely heated areas. The air heats then expands, rises, and flows outward (diverges) at higher levels. Thus, there is a net decrease in the vertical air column. The pressure drops, and the surrounding air rushes in, spiraling inward in a counterclockwise fashion. This type of low lasts most of the summer in southwestern Arizona and southeastern California and in the deserts of the world. Dust devils are very small-scale heat lows (or thermal lows).

Sometimes, as shown in figure 5–12, lows form on the leeward side of a mountain range, where they may cause very annoying weather disturbances. Favorite areas are to the east of the Rockies in Colorado, in the Texas Panhandle, east of the Appalachians, and most areas downwind from almost any mountain range with a height greater than 3,000 feet.

HIGH AND LOW SUMMARY

	HIGHS	LOWS
Weather	Generally fair. Sometimes shallow fog near center.	Usually moderate to heavy rain or snow. Stormy.
Clouds	Usually only in the periphery. Stratocumulus or cumulus in the E, and cirrus in the W.	Almost all types, ranging from high to low.
Circulation	Spiraling outward in a clockwise fashion in northern hemisphere. Spiraling outward in a counterclockwise fashion in southern hemisphere.	Spiraling inward in a counterclockwise fashion in northern hemisphere. Spiraling inward in a clockwise fashion in southern hemisphere.
Winds	Light near center, increasing going outward toward periphery.	Strong everywhere, except in the very center.
Temperature	Warm or cold for relatively long periods of time, with little or no change.	Usually cold, or warm changing to cold. If of tropical origin, very warm.
Average Movement	Winter: 23.5 knots. Summer: 16.0 knots.	Winter: 25.0 knots. Summer: 18.0 knots.
Size	From 200 to 2,000 miles.	From 20 to 2,000 miles.
Shape	Circular or elliptical, and in between.	Circular or elliptical, and in between.
Marine Threat	None, except for possibility of fog near center or in periphery where warm air travels over cold water.	Yes, strong winds and high seas.

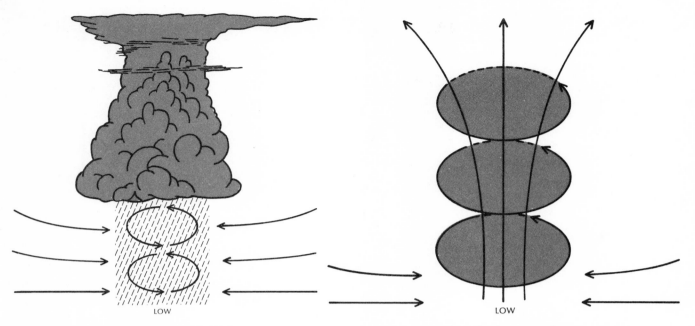

Figure 5–10.   Low formed when air under a large cumulo-nimbus cloud is forced to rise rapidly.

Figure 5–11.   Heat low develops over deserts and other intensely heated places as heated air expands, rises, and flows outward at higher levels.

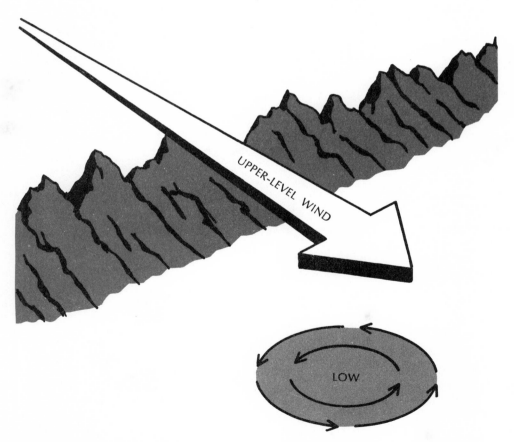

Figure 5–12.   Lows form on the leeward side of mountain ranges where they may cause trouble.

Chapter Six

Air Masses

The term *air mass* is in constant use by weather-men and means *high-pressure cell,* but this term has a much different emphasis when we talk about anticyclones. By definition, an air mass is a vast volume of air, often enveloping many thousands of cubic miles, in which the conditions of temperature, moisture, etc., are much the same at all points in a horizontal direction. An air mass (or high, or anticyclone) takes on the temperature and moisture characteristics of the surface over which it forms. For example, an air mass which forms over the Arctic or northern Canada during the winter months is very cold and dry because of the low temperatures and humidity in those areas.

Almost all continents, including the United States, are swept by air masses of large contrast. The North American continent is rather wide at the top. Consequently, cold and dry air masses form there continually and travel southward. The southern part of the continent is quite narrow. Thus, the moist and hot tropical air masses that form over the oceans can easily move northward. When a hot, moist air mass traveling northward meets a cold, dry air mass traveling southward, a weather *front* usually develops where these two air masses of different origin bump into each other. (See figure 6–1.) The formation of weather fronts will be discussed in detail in the next chapter.

DESIGNATION AND TRANSFORMATION OF AIR MASSES

Air masses traveling around the earth and affecting its weather carry with them the temperature and moisture characteristics of their origin. As they move across oceans and continents, they are modified by the surface over which they travel, but their original characteristics tend to persist.

Air masses usually come from one of two sources —the polar regions or the tropical regions. On weather maps polar air masses are labeled *P;* tropical air masses, *T.* The two types of surfaces over which these air masses are formed are the continental, labeled *c,* and the oceanic, or maritime, labeled *m.* Thus, polar air formed over the oceans is labeled *mP* on a weather map. Polar air formed over a continent is labeled *cP.* A tropical air mass formed over the ocean is labeled *mT.* A third letter is added to indicate whether the air mass is colder (*k*) or warmer (*w*) than the surface over which it is traveling. This is most important to know in order to determine what kind of future weather is in store. So, if continental polar air were moving over a surface which is warmer than itself, it would be labeled *cPk.* If maritime tropical air were traveling over ground that is colder than itself, it would be labeled *mTw.*

There are several different air-mass classification systems in use by weathermen around the world, but we don't have to concern ourselves about these. If we remember the system explained above, we will have an insight into the current and future weather conditions and effects.

Figure 6–2 depicts some of the principal air masses that affect the United States and its offshore areas, and also some of the typical paths which the air masses travel along. Note that the air masses starting in the north and traveling generally southward are labeled *polar.* Also, note that the air masses starting in the south and traveling generally northward are labeled *tropical.*

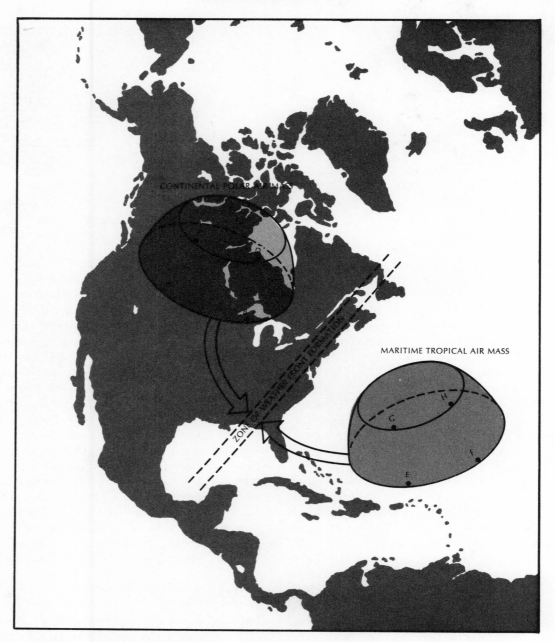

Figure 6–1. Air mass properties, such as temperature and humidity, are very similar at any level. Points of similarity are A and B; C and D; E and F; and G and H.

MARITIME AND CONTINENTAL AIR MASSES

Maritime and continental air masses differ considerably in temperature and moisture; consequently, the weather associated with each is quite dissimilar. The ocean surface and the land surface reflect solar radiation differently. Much more heat is needed to raise water temperature than to raise land temperature. Only the few top inches of solid earth absorb radiation, while the oceans absorb heat to a depth of more than 80 feet. Also, the ocean turbulence carries heat to much greater depths.

The result of this is that the oceans are much slower to warm up and cool down than are the continents. This is the reason why the oceans lag behind the continents by one or two months in their response to seasonal changes. Believe it or not, there is less than an 18° F difference between an average winter and summer temperature over most open ocean areas. This thermal lag and its moderating effects arise because a great deal of spring heat on the ocean is used in melting ice in the Arctic and

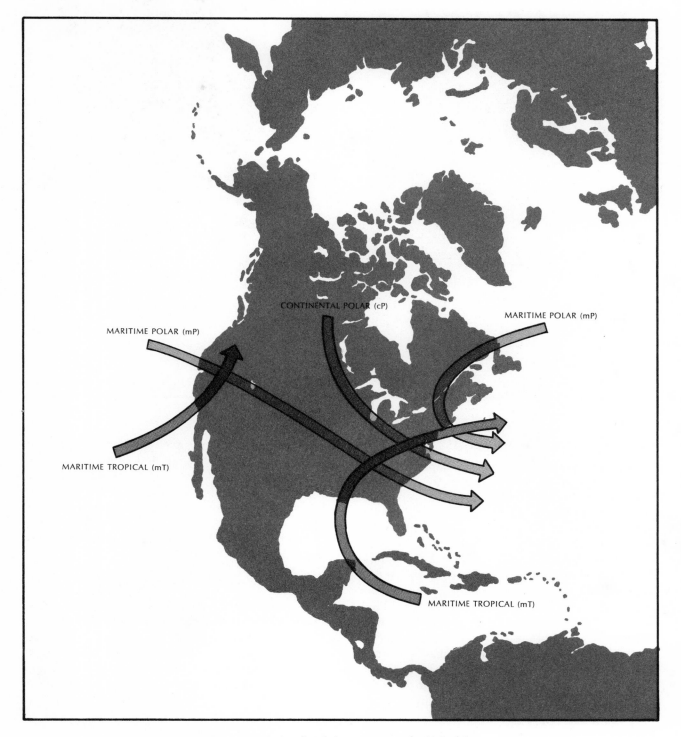

Figure 6–2. Typical paths of air masses over the United States

Antarctic Oceans. Just as much heat is required to melt a pound of ice at 32° F to water at 32° F as is required to heat a pound·of water from 32° F to 176° F. Consequently, the oceans warm very slowly in the spring and cool slowly in the fall because the same amount of heat is thrown off as the Arctic water freezes. This heat warms the air and causes a lag in the approach of cold weather. Maritime air tends, therefore, to moderate extremes of either cold or heat as it moves over the continents.

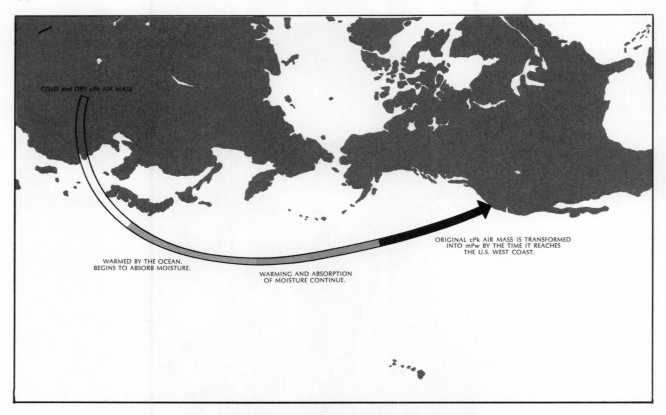

Figure 6–3.   Changes in polar air masses crossing the Pacific

POLAR AND TROPICAL AIR MASSES

Polar air masses usually change rapidly as they sweep southward and eastward over the continents. Since cP air is initially colder (cPk) than the land surface over which it travels, the bottom layers of air are warmed by the earth's surface and forced to rise as vertical currents, creating turbulence, cumulus-type clouds, and rain or snow in the form of showers. The air is further warmed by the heat of condensation (as the water vapor condenses and releases the heat of condensation to the air). Thus, polar air tends to "mix" and become somewhat warmed. For example, as shown in figure 6–3, cP air starting out over Siberia as a dry air mass is warmed and moistened as it travels across the Pacific Ocean, acquiring the characteristics of mP air. This air, which has been gradually changed from cP to mP air in its travel across the Pacific, is cooled as it passes over the west coast of the United States and over the mountains, causing it to drop its moisture (as rain or snow). It then continues its eastward travel as a relatively dry and warm air mass.

Tropical air masses, on the other hand, change much more slowly because they are warmer than the land over which they travel. The bottom layers of tropical air masses are cooled by the land sur-face over which they travel, a stabilizing process. There is very little mixing, and continued cooling of this air mass (mT) as it travels northward tends to increase this stability. As a result, mT air tends to remain the same for very long periods of time. For example, the hot, humid days during the summer months in the middle west and on the east coast seem to go on forever.

AIR MASS STABILITY

The stability of cold and warm air masses determines what sort of weather these air masses bring along with them. Vertical air currents are caused by the heating and cooling of air by contact with the earth's surface and by the flow of air over and down mountain barriers. Stable air resists these vertical currents and returns quickly to normal. In unstable air, the vertical currents are allowed to grow and accelerate.

An air mass traveling over a surface that is colder than itself is *thermodynamically warm*. Such an air mass is termed *stable*; it resists vertical motion. On the other hand, an air mass moving over a surface which is warmer than itself is thermodynamically cold and is generally unstable. Vertical motion grows or is accelerated.

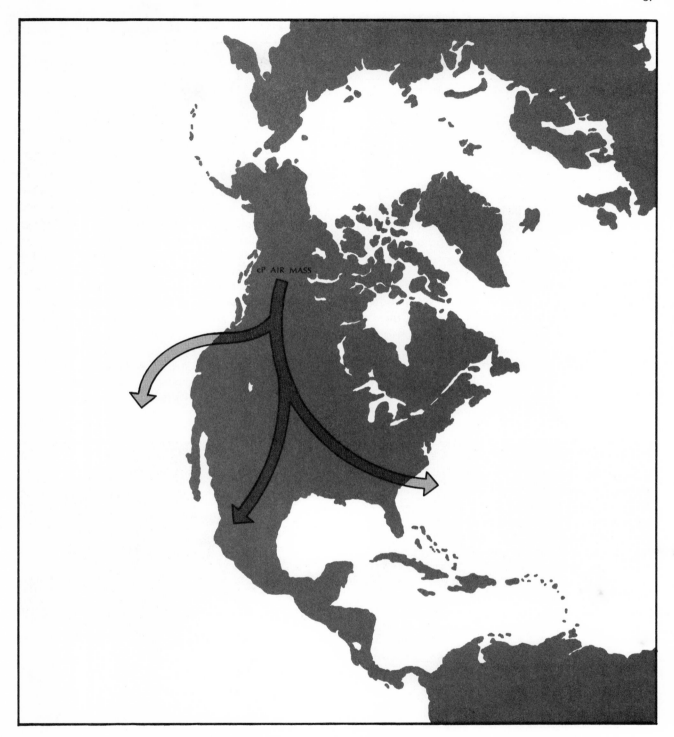

Figure 6–4. Common paths of continental polar (cP) air masses in winter

A thermodynamically warm air mass is cooled by contact with the colder earth's surface below. The cooled, heavy, lower surface air tends to stay at the bottom. If it is lifted by one means or another, it sinks back to its original level. But when a thermodynamically cold air mass is warmed by contact with the earth's surface, the heated, light air rises through the cold air above. Consequently, a thermodynamically cold air mass (unstable) is turbulent, with cumulus-type clouds and showery conditions. Table 6–1 is a comparison of cold and warm air masses for ready reference.

TABLE 6–1   Weather Features of Cold (k) and Warm (w) Air Masses

Air Mass Suffix	Stability	Turbulence	Surface Wind	Visibility	Clouds	Precipitation
Cold (k)	Unstable	Yes	Gusty	Good	Cumulus-Type	Showers or Thundershowers
Warm (w)	Stable	No	Steady	Poor	Stratus-Type	Drizzle

AIR MASSES OF WINTER

For purposes of air mass discussion and illustration, concentration is placed upon the geograpical area of the United States and the waters contiguous thereto. However, the same general principles apply to air masses all around the globe.

Continental Polar (cP)

Continental polar air masses are extremely cold. When they travel over warmer surfaces, they are designated *continental polar cold* (*cPk*). Since these air masses are colder than the underlying surface, they are highly unstable (turbulent). Their most common paths are as shown in figure 6–4. The most common path is marked X. As the *cPk* air travels southward over the Great Lakes—especially before they become frozen—the lower layers of the air mass are warmed considerably and absorb considerable moisture from the lakes at about 34–39° F. The warmed, moistened lower air-parcels rise through the colder upper air, causing snow flurries or heavy snow showers on the leeward side of the lakes. Localities on the southern and eastern shores of the lakes receive more than five times as much snow in a winter than those on the northern and western shores. As the air mass ascends over the Appalachians, the *cPk* air is cooled further, and more precipitation results on the windward side of the mountains. To the east of the Appalachians, however, the air sinks, is heated, and becomes fairly dry. Consequently, the skies are generally clear on the leeward side.

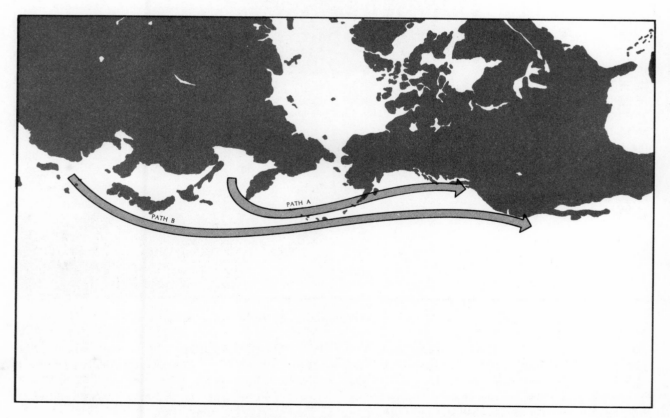

Figure 6–5. Typical paths of maritime polar (mP) air masses in winter

On occasion, the cPk air follows path Y on figure 6–4. The air is still turbulent and cumulus-type clouds form. Since the cPk air neither passes over the warm water of the lakes nor is forced to ascend over mountains, this means generally clearer skies than when the cPk air is traveling along path X. The third path Z, is quite rare. On this path, the cPk air mass moves over the eastern Pacific, is warmed, and collects moisture. The effect is usually squalls and even snow as far south as southern California.

Maritime Polar (mP)

Maritime polar air masses may take a relatively short journey over the Pacific, as shown by path A on figure 6–5, before reaching the coast. Rain or snow showers are common as the air is lifted over the coastal mountain ranges. If the air mass crosses over the Rockies and stagnates in the Mackenzie Valley, it acquires the characteristics of a cP air mass.

If the path of the mP air mass is considerably longer over the ocean surface, as shown in figure 6–5 (path B), it is warmed a great deal more by the ocean surface and absorbs considerably more moisture. This type of mP air mass is very common on the Pacific Coast in winter. The air masses are cooled on contact with the land surface, and rain and fog are the inevitable result.

Maritime polar air masses rising over mountains produce many clouds and moderate to heavy precipitation on windward slopes. Sometimes these air masses stagnate between the coastal mountains and the Rockies in the Great Basin region. If these air masses travel eastward over the Rockies, they are warmed as they descend the east slopes and bring clear skies, low humidity, and generally mild weather with them.

Maritime Tropical (mT)

Maritime tropical air masses are hot and humid at their source regions. Since they are much warmer than the land surfaces over which they travel, the air masses are stable and bear the title mTw. Stratus or stratocumulus clouds often form, mostly as the result of nighttime cooling. These tend to disappear by about noon because of the warming which takes place. Little or no precipitation occurs unless the mTw air is forced to ascend over a wedge of cold air, which acts just like a mountain range. In this case, a steady type of rain results. This happens quite often and accounts for a great deal of the rain in the United States.

Maritime tropical air masses seldom move over the north or northeast parts of the United States in winter. When they do, the mTw air is rapidly cooled by the cold land surface and extensive fog results. These air masses also cause dense ocean fogs (to be discussed in chapter 9) as they travel from the Gulf Stream area toward the cold waters of the North Atlantic. Figure 6–6 illustrates the more common paths of mT air masses in winter.

Maritime tropical air masses rarely enter the United States via the Pacific Coast. This happens only when extremely low pressure exists off the California coast. In this case, the warm and moist mT air crosses the coastline and is forced to rise rapidly over the much colder and heavier air that it encounters, or over the Sierra Nevada and other western ranges. When this happens, and despite the fact that the mT air is very stable, it produces moderate to heavy rainfall in southern California.

AIR MASSES OF SUMMER

Continental Polar (cP)

Continental polar air masses are quite different in summer than in winter. The source regions are much warmer, and the general circulation of the earth's atmosphere is much weaker because there is less of a temperature difference between the polar and tropical regions. The air moves at a congenial, rather than at a frantic, pace. Generally fair weather prevails, but a few cumulus-type clouds are produced by local heating. Rain occurs over the Great Lakes and on the windward slopes of mountains.

Maritime Polar (mP)

On the Atlantic side, maritime polar air masses form over the cold waters of the North Atlantic. These air masses are colder than the cP or mT air masses which generally sweep the eastern states. Occasionally, mP air masses flow in over the land. When they do, they bring with them the "cool breaks" which sometimes may reach as far south as Florida. If an mP air mass makes contact with an mT air mass from the south, frontal weather (low stratus clouds and drizzle) occurs as the mT air mass is lifted over the cooler mP air and is cooled adiabatically. This process will be discussed in chapter 7.

On the Pacific side, the mP air masses in summer produce the sea and coastal fogs so famous in the California area. The moist mP air masses pass over the cold ocean surface off the coast. As they are cooled from below, the mP air produces dense sea fog, which rolls inland, and may last for weeks. Further inland, the *mP* air is warmed quickly by contact with the land surface, and the relative humidity drops abruptly. Although the air mass is unstable, the relative humidity is lowered to such an extent inland that skies are clear from areas just inside the coastline to the Sierras. But some shower activity may occur on the west slopes.

The cold ocean current that causes the west coast fogs is the result of "upwelling" of the subsurface

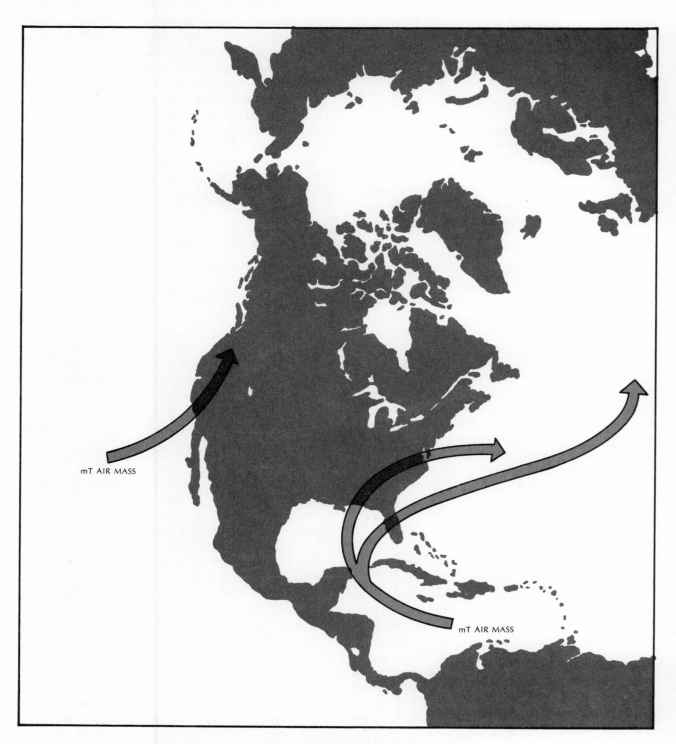

Figure 6–7. Typical paths of maritime tropical (mT) air masses in winter

waters. The friction of the prevailing northwest wind over the Pacific pushes the coastal surface waters to the right (i.e., toward the southwest). The result is that the colder subsurface water rises, or wells, up. This colder water cools the air mass above it, producing the fog and stratus clouds which may then move in over the coastal region, producing cool weather and raising havoc with one's plans.

Maritime Tropical (mT)

On the Atlantic side, the mT air masses are probably the most important of all air masses during the summer months. These are the air masses in which nature brews the dreaded hurricanes. Also, the central and eastern states (and sometimes the southwest) are greatly affected by these air masses. The Atlantic mT air travels almost continuously over the central and eastern states during the summer, bringing with it excessive humidity and oppressive heat.

The mT air masses formed in the Caribbean are extremely hot and humid. They are very unstable as they move in over the hotter southern coastal states. Since the land surface cools quite rapidly at night, the air mass's stability increases and stratocumulus clouds tend to form, with their bases near the ground. But these clouds usually clear by midmorning and the land surface again warms and heats the air. Further heating later in the day increases the instability again, and showers and thunderstorms may result from the rapidly rising and moist air which is cooled during ascent to a point below its saturation point.

On the Pacific side, mT air masses in summer are of little importance. They do, on occasion, travel inland and may reach as far north as Alaska. But the air masses are almost always mTw (warmer than the underlying surface), so they are stable and cause little or no weather threat.

Chapter Seven

Weather Fronts and Frontal Weather

AIR-MASS BOUNDARIES—FRONTS

As we discussed in the previous chapter, although each individual air mass is quite homogeneous in a horizontal direction (at almost all levels), the boundary zones of *different* air masses can be very sharp. This is because the mixing of air across the boundary zone between two different and distinct air masses having different temperature and humidity conditions takes place very, very slowly. On weather maps, the boundary zones of the different air masses are drawn in as lines, which are called *weather fronts*, or simply *fronts*. Thus, a *front* may be regarded as a line at the earth's surface, or any other nearly horizontal surface, dividing two different air masses.

The boundaries between air masses are really *zones of transition*, ranging from about 5 to 60 miles in width, but because of the small geographic scale of weather maps, fronts are usually drawn as lines on the meteorological charts.

If we exercise our mental processes and think in three dimensions, we can visualize two dissimilar air masses (as in figure 6–1) coming together and being separated by a surface whose intersection with any horizontal plane is the *front* as represented on that particular chart. This surface of separation between the two air masses is known as a *frontal surface*. The name *front* was first introduced during World War I by analogy with battlefronts. But the analogy goes much further, because most weather disturbances originate at fronts, and the general weather picture, as successive disturbances travel along the frontal zone, is one of constant war between two or more conflicting air masses.

The principal frontal zones around the earth for summer and winter are as shown in figure 7–1.

Arctic, separating Arctic (A) air masses from either maritime polar (mP) or continental polar (cP) air masses.

Polar, separating continental polar (cP) air masses from maritime polar (mP) air masses, or separating either cP or mP air masses from maritime tropical (mT) air masses.

Mediterranean, separating the cold air masses over Europe during the winter months from the warmer air masses over North Africa.

Intertropical Convergence (*ITCZ*), referred to in older publications as either the *intertropical front,* or the *equatorial front.* This is the normally bad-weather region in tropical latitudes where the northeast trade winds of the northern hemisphere and the southeast trades of the southern hemisphere either approach each other at a rather large angle, or flow almost parallel to each other. The ITCZ is not really a true front, like the fronts of middle or high latitudes, because the temperature and moisture conditions on both sides of the ITCZ are very nearly the same. This zone is one of the favorite breeding areas of hurricanes and typhoons, which will be discussed in chapter 8.

73

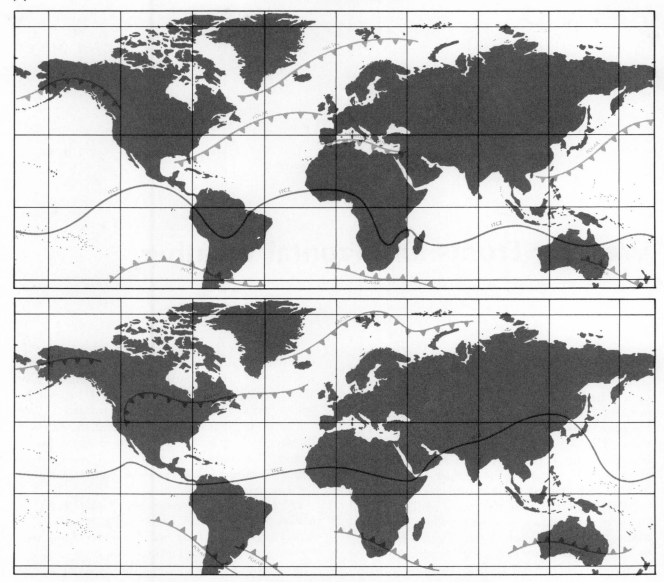

Figure 7–1.   Mean positions of frontal zones in summer and winter

SOME FACTS ABOUT FRONTS

Figure 7–2 is a plan view of a front, as though you are an astronaut, looking down from high above. A cold anticyclone is to the left (west), and a warm anticyclone is to the right (east). The arrows show the wind directions. The isobars show the pressure in millibars. The direction of movement of the cold front is from west to east (i.e. at the earth's surface, cold air is replacing warm air). Figure 7–3 is a block view of the same cold front, and the line *X-Y* has the same position in each diagram.

After studying these diagrams, certain facts, true of almost all fronts, should be remembered:

1. Fronts form at the outer boundaries of high-pressure cells.

2. Fronts form only between air masses of different temperature and/or moisture conditions.

3. The warm air always slopes upward over cold air because it is less dense (lighter).

4. Fronts are found along low-pressure troughs, with few exceptions. Consequently, the pressure falls as a front approaches and rises after it passes.

5. The wind near the ground always shifts clockwise (in the northern hemisphere) as a front passes over your position.

6. The isobars at a front are always V-shaped.

7. A front always slopes upward over the cold air either ahead of, or to the rear of, the front's direction of advance.

8. There are four basic types of fronts—cold, warm, occluded, and stationary.

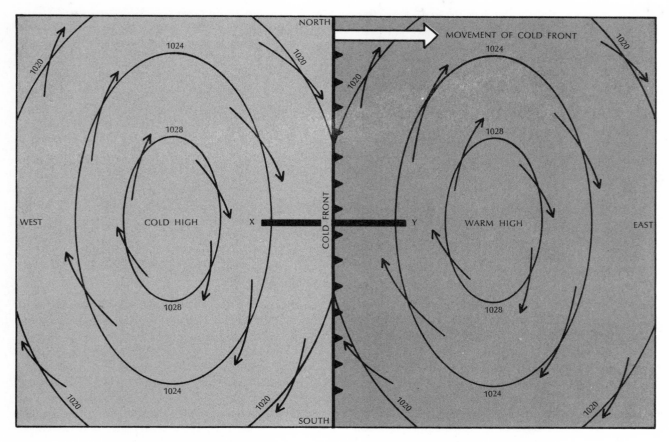

Figure 7–2.   Plan view of a cold front and its associated cold and warm anticyclones

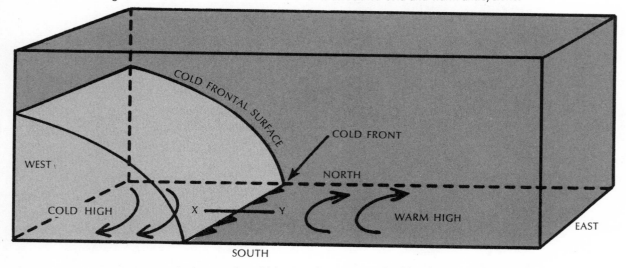

Figure 7–3.   Block view of a cold front and its associated cold and warm anticyclones

COLD FRONTS

Referring again to figure 7–3, we see that cold fronts wedge their way *under* warm air (since the cold air is heavier) as they advance. The typical thick wedge of a cold front develops, as shown in figures 7–4 and 7–5, because the friction with the ground tends to hold back the very bottom of the advancing cold air mass. As a result, the cold air tends to pile up into a rounded prow-shape as it advances against the warm air.

In the northern hemisphere, cold fronts usually are oriented in a northeast-to-southwest direction and move toward the east or southeast. Since the general movement is toward the east, cold air masses advance in that direction. Remember that cold fronts are the easterly and southerly edges of cold air masses.

The speed of cold fronts varies within rather wide limits (10 to 50 knots), but as a general rule, the speed is considerably lower in summer than in winter. This is because in winter the air is much colder and exerts greater pressure. On the average, cold fronts during the summer months travel at speeds about ⅓ to ½ of the winter speeds.

Although the sloping edge of a cold front may extend over several hundred miles horizontally, the steepness of the advancing edge (a ratio of about 1/50) means that the frontal weather is limited to an extremely narrow band. See figures 7–4 and 7–5, which are necessarily exaggerated horizontally. The sharply sloping edge also produces an abrupt lifting of the warm air, so that the storms associated with cold fronts are usually violent, though relatively brief.

Weather at slowly moving cold fronts differs quite a bit from the weather accompanying fast-moving cold fronts. If the warm air ahead of the cold front is stable, nimbostratus clouds will form almost di-rectly over the cold front's contact with the ground, and rain will fall through the cold air mass after the front has passed your position. If the cold front is very weak, only a few clouds will form and chances are that there will be no rain (figure 7–4).

On the other hand, if the warm air ahead of the cold front is unstable and very humid, cumulonim-bus clouds will form and heavy thundershowers will occur as the warm, humid air is lifted over the wedge of cold air. The main rainfall, however, will occur in the cold air mass after the front has passed over your position. Typically, a steady downpour from the nimbostratus clouds at the lower levels alternates with the heavy showers from the cu-mulonimbus clouds towering to great heights. (See figure 7–5.)

Squall Lines

Squall lines frequently precede fast-moving cold fronts. They are an almost unbroken line of ominous, black clouds, sometimes towering to 40,000 feet or higher, including thunderstorms of great violence and sometimes tornadoes. These phenomena are usually extremely turbulent. They can inflict con-siderable damage to small boats and should be avoided if at all possible. From a boat, a squall line

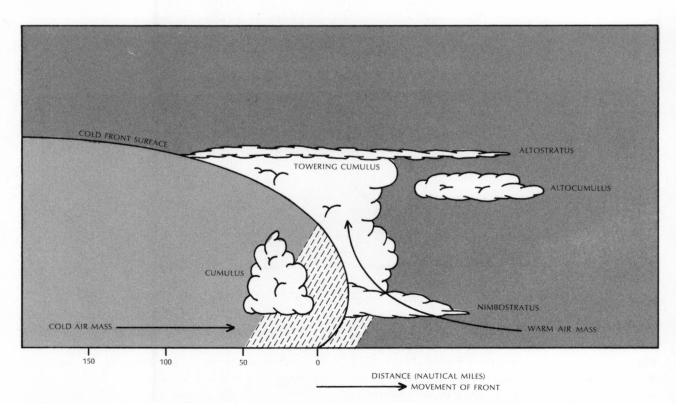

Figure 7–4.   Vertical cross section of a slow-moving cold front lifting stable air

looks like a veritable wall of rolling, boiling, black fog. The wind shifts and increases suddenly with the approach of a squall line, and the deluge of rain may carry the cloud right down to the ground in sharp, vertical bands. Torrential rain falls behind the leading edge of the squall line.

Squall lines normally occur when the winds above a cold front, moving in the same direction in which the front is advancing, prevent the lifting of the warm air mass by the wedge of the cold air mass. In this case, very little bad weather occurs ahead of the advancing cold front. But 100 to 150 miles ahead of the cold front, the strong winds force up the warm air with almost explosive violence, producing the dreaded and hazardous squall line.

Cold Front Weather Sequence

The sequence of weather associated with a cold front is usually the same. First, one notices an increase of winds from the south or southwest, and the appearance of altocumulus clouds darkening the horizon to the northwest or west. Also, the barometer begins to fall.

As the cold front approaches, the clouds lower and cumulonimbus clouds begin to tower in the sky. Rain begins to fall and increases in intensity quite rapidly. The wind usually increases, and the barometer falls still further.

With the actual passage of the cold front over your position, the wind shifts rapidly to the west or northwest, with strong gusts. Squally weather continues, and the barometer reaches its lowest reading. Passage of the front usually results in fairly rapid clearing, but in mountainous areas, cumulus or stratocumulus clouds may persist for some time. The barometer begins to rise rapidly, and the temperature drops. Then winds generally become rather steady from the west or northwest.

After the passage of a cold front, a few days of typically good "anticyclone weather" are usually in store.

WARM FRONTS

Warm fronts are those fronts behind which warm air advances and replaces cold air at the earth's surface (or at levels above the earth's surface). In the northern hemisphere, warm fronts usually occur on the east side of low-pressure systems and are usually followed by cold fronts as the prevailing westerly winds push the low eastward. See figures 7–6 and 7–7 for a plan view and a block view of the warm front phenomenon.

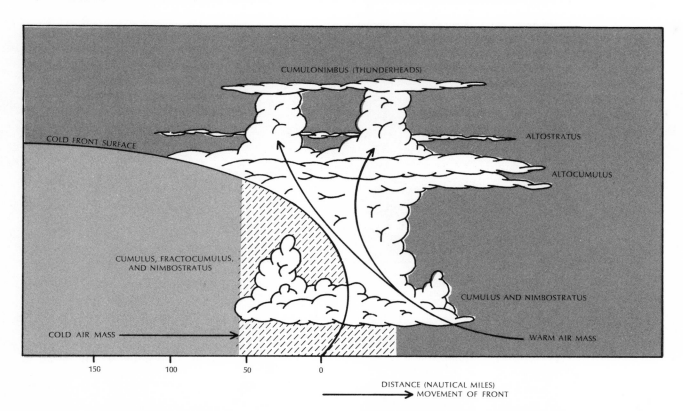

Figure 7–5. Vertical cross section of a slow-moving cold front lifting unstable warm air

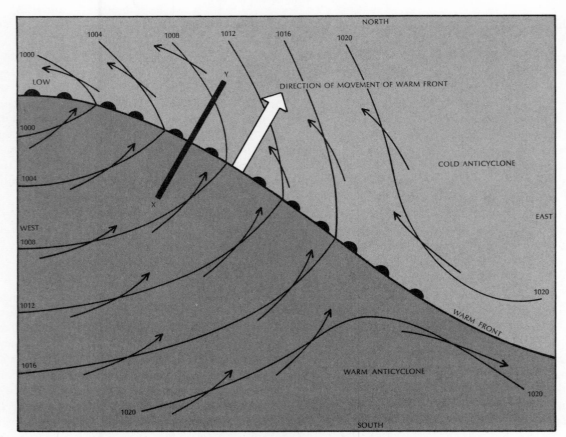

Figure 7–6.   Plan view of a warm front

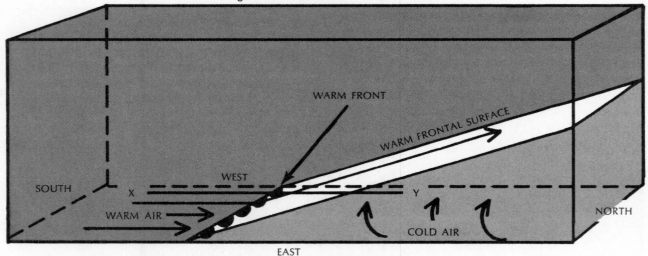

Figure 7–7.   Three-dimensional view of a warm front. The warm air of the warm high is riding
up over the wedge of cold air associated with the cold high.

The speed of advance of warm fronts in a horizontal direction is about half the speed of cold
fronts. On the average, they travel at about 12 to 15
knots. The vertical slope of a warm frontal surface
is much less than that of a cold front, with a ratio
averaging between 1/100 to 1/200 (whereas the
slope of a cold front is 1/50). The warm air moves
rather gradually up the slope, which does *not* have
the typical cold-front bulge. This is because the
friction at the ground drags the bottom edge of
the retreating cold air ahead of the warm front
into a thin wedge.

Warm-front weather extends over an area hundreds of miles in advance of the front line at the

ground. Typical cloud sequences can be seen as far as 1,000 miles in advance of the front and often 48 hours before the front's arrival at your position. The clouds and precipitation typical of warm fronts develop along the contact zone of the warm and cold air masses above the ground.

Stable and Unstable Warm Air

As shown in figure 7–8, when *stable* warm air is lifted over cold air as the warm front advances, stratus-type clouds are produced because the uplift of the warm air is slow and gradual, and there is very little turbulence in the warm, stable air. As the warm air is lifted, it cools to produce clouds in the following order: stratus, nimbostratus, alto-stratus, cirrostratus, and cirrus clouds. Precipitation is heavy at the beginning of the uplift but decreases gradually as the cloud bases become higher, leaving relatively dry cirrus clouds above 20,000 feet, far in advance of the warm front.

As seen in figure 7–9, *unstable* warm air produces more violent weather. The turbulence in the warm air mass is great, and the unstable air produces strong ascending air currents which create cumulonimbus clouds (thunderheads) ahead of the front line. Consequently, the precipitation alternates between heavy downpours and moderate drizzle, with thunderstorms interspersed amongst everything else.

The Rebel Fronts

Sometimes fronts move very little, oscillate, or do not move at all. These are called *stationary* fronts by weathermen. Weather conditions are very much like those associated with warm fronts but are usually more mild. Stationary fronts with rain or snow may hang on for days at a time, much to everyone's frustration.

Weak fronts sometimes will pass your position, unnoticed by everyone except the meteorologists. These fronts occur when the two air masses are almost identical in their temperature and moisture conditions and are marked by only a wind shift as the front passes. Weathermen keep close tabs on weak fronts because they sometimes regenerate into strong fronts.

Warm Front Weather Sequence

The sequence of weather associated with warm fronts is one of the most reliable displays in the weather business. The first sign is almost always the

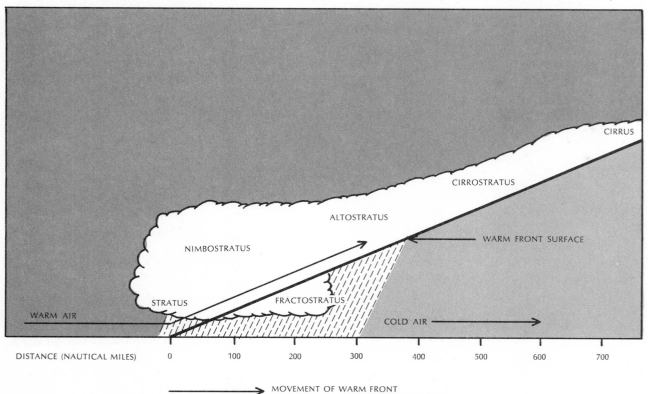

Figure 7–8. Vertical cross section of a warm front with stable warm air

appearance of the high cirrus clouds which have been lifted the farthest up the cold-air slope, way in advance of the front. These change to cirrostratus clouds, which become more dense as the front continues to advance. Also, the barometer begins to fall.

If cirrocumulus clouds (mackerel sky) appear, this means that the warm air aloft is unstable and the situation is as shown in figure 7–9. Heavy downpours, moderate drizzle, and thunderstorms should be expected. Just about every experienced mariner and farmer knows this. The barometer continues to fall.

If the warm air overhead is stable, the cirrus and cirrostratus clouds are replaced by lower altostratus clouds. Rain or snow begins as the altostratus overcast becomes more dense, and the precipitation continues until the warm front has passed your position. The barometer continues to fall until the front reaches you and then levels off.

Nimbostratus, stratus, and stratocumulus clouds (and in unstable air, cumulonimbus clouds) complete the warm-front sequence. At this point, we should refresh our memories by taking another look at the cloud photographs in chapter 3. The rain falling through the wedge of cold air adds to the moisture content of the cold air, and this produces stratus clouds, which frequently obscure the higher clouds. As the warm front passes, the skies clear, the temperature rises, the barometer rises, and the wind shifts clockwise (usually from southeast to southwest) in the northern hemisphere.

OCCLUDED FRONTS

Occluded fronts include both the best and the worst features of cold fronts and warm fronts. An *occluded front* is generated when a cold front—following closely on the heels of, and moving faster than, a warm front—finally overtakes the warm front and lifts the warm air mass completely off the ground. Either the cold frontal surface or the warm frontal surface is forced to rise off the ground, depending upon whether the cold air behind the cold front is colder or warmer than the cold air in advance of the warm front.

Figure 7–10 shows the occlusion process in vertical cross-section. In this case, the cold air behind the cold front is colder (and heavier) than the cold air ahead of the warm front. The result is a cold-front type of occlusion, wherein the warm front surface is lifted aloft. Figure 7–11 is a plan view of the three fronts after occlusion has taken place.

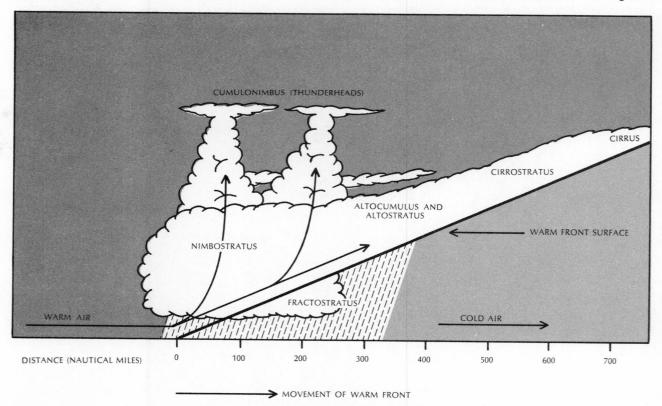

Figure 7–9. Vertical cross section of a warm front with unstable warm air

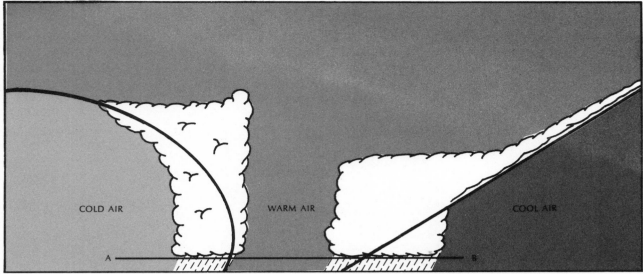

(a) THE COLD FRONT APPROACHING THE WARM FRONT, PRIOR TO OCCLUSION

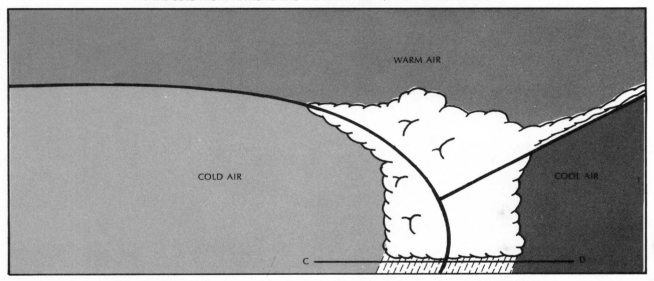

(b) OCCLUSION OCCURS AS THE COLD FRONT OVERTAKES THE WARM FRONT AND LIFTS BOTH THE WARM FRONT AND THE WARM AIR MASS COMPLETELY OFF THE GROUND.

Figure 7–10. Vertical cross section of the occlusion process (all fronts are moving from left to right)

The lines *AB* and *CD* illustrate the situation before and after occlusion takes place, respectively, in both vertical and horizontal planes. Though not colored in the illustration, blue is used for the cold front surface, red for the warm front surface, and purple for the occluded front surface. This is the color scheme of all weathermen.

Figure 7–12 illustrates the case where the cold air ahead of the warm front is colder (and heavier) than the cold air behind the cold front. The result is a warm-front type of occlusion, wherein the cold-front surface is lifted aloft. In the case of occluded fronts, whether cold-front or warm-front type, precipitation occurs on both sides of the front. The

cloud and weather sequences of occluded fronts are as already described for warm and cold fronts.

MEMOIRS OF A FRONTAL LOW

Fronts around the world tend to move in a general west-to-east direction. Sometimes there is a large northerly or southerly component to the general easterly movement, and sometimes they stand still or move backward (called *retrogression*). For ready reference and convenience, tables 7–1 and 7–2 contain summaries in quick-glance form of the major weather elements and their changes involved with cold fronts and warm fronts, respectively.

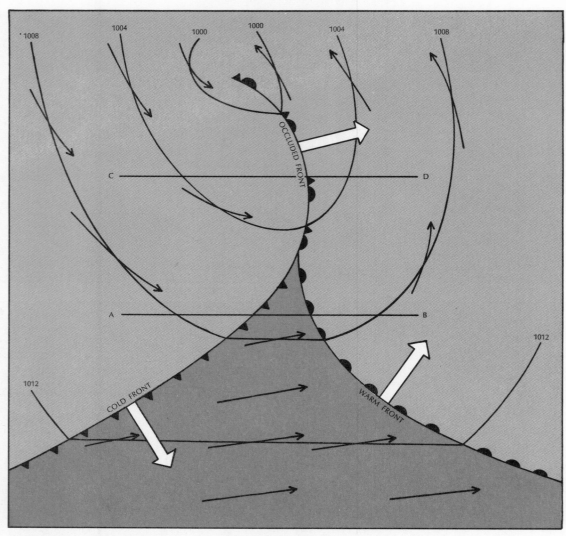

Figure 7–11.   Plan view of the occlusion process. The lines *AB* and *CD* correspond to the same
lines on the vertical cross sections in figure 7–10.

Table 7–1   Summary of Weather Sequence at a Cold Front

Element	In Advance	During Passage	To the Rear
WEATHER	Usually some rain; perhaps thunder.	Heavy rain; perhaps thunder & hail.	Heavy rain for short period, then fair. Perhaps scattered showers.
CLOUDS	Ac or As & Ns, then heavy Cb.	Cb with Fs or scud.	Lifting rapidly, followed by As or Ac; perhaps Cu later.
WINDS (northern hemisphere)	Increasing and becoming squally.	Sudden clockwise shift; very squally.	Gusty.
PRESSURE	Moderate to rapid falls.	Sudden rise.	Rise continue more slowly.
TEMPERATURE	Fairly steady; may drop a bit in pre-frontal rain.	Sudden drop.	Continued slow drop.
VISIBILITY	Usually poor.	Temporarily poor, followed by rapid improvement.	Usually very good, except in scattered showers.

Table 7–2   Summary of Weather Sequence at a Warm Front

Element	In Advance	During Passage	To the Rear
WEATHER	Continuous rain or snow.	Precipitation usually stops.	Sometimes a light drizzle or fine rain.
CLOUDS	In succession: Ci, Cs, As, Ns; sometimes Cb.	Low Ns and scud.	St or Sc, sometimes Cb.
WINDS (northern hemisphere)	Increasing.	Clockwise shift, sometimes decreasing.	Steady direction.
PRESSURE	Steady fall.	Levels off.	Little change; perhaps slight rise followed by slight fall.
TEMPERATURE	Steady or slow rise.	Steady rise, usually not sudden.	Little change or very slow rise.
VISIBILITY	Fairly good except in precipitation.	Poor; often mist or fog.	Fair or poor; mist or fog may persist.

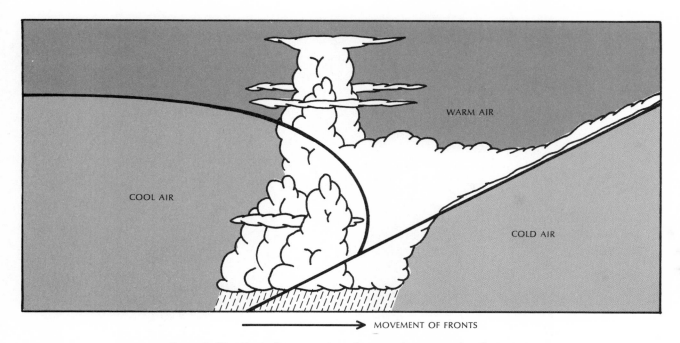

Figure 7–12.   Vertical cross section of a warm front type of occlusion

Warm fronts are usually followed rather closely by cold fronts. To understand why, let's consider the way in which nontropical (also called extra-tropical) lows develop from wave-like outbreaks along frontal lines. Figure 7–13 (a) through (f) illustrates the development of a typical low-pressure cell as you would see it from a space capsule orbiting the earth. Remember, the black lines are isobars; light arrows show the flow of cold air, and heavy arrows warm air.

In figure 7–13(a), a stationary front exists in a trough of low pressure between a cold and a warm air mass. As can be seen in figure 7–13(b), the cold air begins to push under the warm air at some point, and a wave-like pattern begins to develop along the front. In figure 7–13(c), the cold front continues to push back the warm air and slides under it. The warm air, pushed from one side, bulges out at its front into a low-pressure area, which increases in size as the circulation intensifies. Consequently, the warm front advances, adding to the development of the wave. In figure 7–13(d), the cold air plunges ahead about twice as fast as the warm front. Consequently, the cold air leaves an ever-expanding low-pressure area behind it, and the wave forms a distinct crest. The cold front finally overtakes the warm front in figure 7–13(e), lifting the warm air completely off the ground and resulting in the formation of the occluded front. In figure 7–13(f), only the low-pressure cell of counterclockwise, inward-spiraling air remains near the earth's surface. Eventually, the swirling low-pressure cell disappears as the air pressure equalizes. Finally, a new front

line is established at the boundaries of the cold and warm air masses. And the process is repeated.

The development and disappearance of low-pressure cells does not take place in a stationary position relative to the earth. The prevailing westerly winds move the air masses in a general southwest to northeast direction. There are exceptions, of course. Thus, as a low develops and a cold air mass bends around the "tongue" of a warm air mass, the warm fronts move along, followed closely by the cold fronts, in a seemingly endless procession.

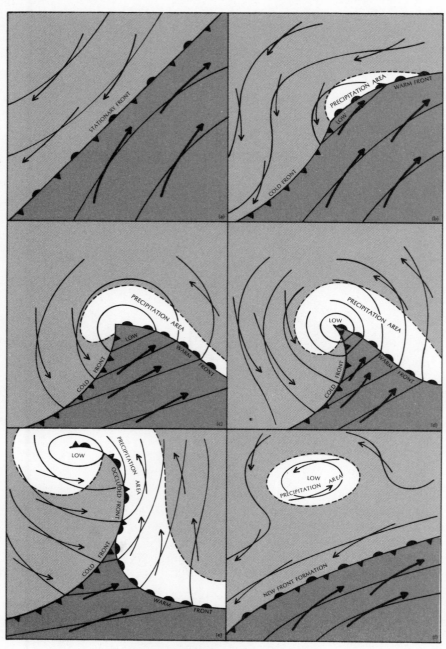

Figure 7–13.   Life history of a frontal low

Chapter Eight

Weather Disturbances and Storms

"Never mind the weather, just so the wind don't blow!" goes an old song that has been sung by mariners for many decades. But the winds *do* blow. They blow comfort when gentle, annoyance when strong, and destruction when violent. In both the tropics and in middle and high latitudes, the winds swirl over the earth in various patterns—some dangerous, some non-hazardous, and some rather pleasant. As L. F. Richardson put it:

> *Big whirls have little whirls*
> *That feed on their velocity,*
> *And little whirls have lesser whirls*
> *And so on to viscosity.*

IN THE TROPICS

Compared to the usually rapid—and sometimes violent—weather changes associated with mid- and high-latitude low-pressure systems, the weather in the tropics follows a rather routine schedule most of the time. The frequent seesawing between cold north winds and warm south winds, accompanied by large temperature falls and rises, is missing. Instead, temperature and wind experience a daily cycle which is dictated largely by orographic, coastal, island, or other terrain features.

In some areas of the tropics, particularly in the large trade-wind regions, this dull, daily cycle so dominates the weather that nothing unusual ever seems to occur—except for hurricanes and typhoons. But this view is grossly exaggerated—especially from a weatherman's standpoint. In tropical regions, a cycle of wet and dry seasons replaces the four seasons of middle and higher latitudes, which are determined by temperature. There is a definite sequence of weather changes during the wet (summer) season and sometimes also during the dry (winter) season.

It is a well-known fact that disturbances of one kind or another in the tropics produce more than 90 percent of the rainfall in this region and that a large fraction of the rain falls in a few intense spurts during the year.

There are many different types of tropical disturbances, which vary largely according to their location. Geographical differences are much larger in the tropics than in higher latitudes, where the basic fundamentals of low-pressure systems hold for all areas and all seasons. In the tropics, many unsolved problems remain. The primary reasons for this are the extreme paucity of weather data in tropical latitudes and the lack of funds and personnel for basic and applied research. For our purposes, then, let's limit the discussion to the more well-known types of disturbances in these latitudes.

INTERTROPICAL CONVERGENCE ZONE (ITCZ)

An almost-continuous trough, or belt, of low pressure at the earth's surface extends around our planet in the equatorial regions. This is where the northeast trade winds (of the northern hemisphere) and the southeast trades (of the southern hemisphere) come together—or converge. This region, within which the northeast and southeast trade winds converge, is known by several different names—the Intertropical Convergence Zone (ITCZ), the Equatorial Trough, the Equatorial Front, the Intertropical Front, the Equatorial Convergence Zone, etc. In this book, we shall refer to it as the *Intertropical Convergence Zone—the ITCZ*.

Seasonally, the ITCZ migrates from the winter hemisphere into the summer hemisphere, so that the winter hemisphere controls a larger fraction of the oceanic tropics than does the summer hemisphere.

Figure 8–1 illustrates the mean positions of the ITCZ during the winter and summer periods of the northern hemisphere. Remember, when it is winter in the northern hemisphere, it is summer in the southern hemisphere, and vice versa.

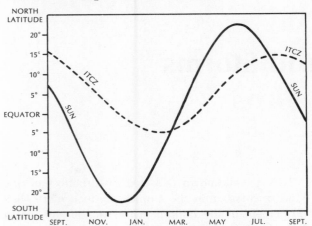

Figure 8–1. Seasonal meandering of the ITCZ compared to the overhead position of the sun. Note the two-month lag of the ITCZ.

In general, as shown in figure 8–1, as the ITCZ migrates north and south in the course of a year, it passes the latitudes between its extreme positions twice, lagging behind the sun by about two months. As a result, many parts of the tropics—but by no means all—have two rainy seasons and two dry seasons. Where the ITCZ moves as shown in figure 8–1, the peaks of the rainy seasons occur in December and April on the equator, in November and May at latitude 5 degrees north, and in June and October at latitude 10 degrees north. At latitudes 15 degrees north and 5 degrees south, the two rainy seasons merge into a single broad peak during the summer months.

The total annual meandering of the ITCZ in the western Pacific sometimes amounts to as much as 44 degrees of latitude in extreme cases. As shown in figure 8–1, the ITCZ reaches its northernmost latitude in August and its southernmost latitude in February. It must be realized, however, that the day-to-day variations in the position of the ITCZ may differ substantially from the mean position during both the winter and summer months.

The ITCZ is usually characterized by strong, ascending air currents, a great deal of cloudiness, and frequent heavy showers and thunderstorms. The intensity does, however, vary greatly. Sometimes the ITCZ looks like a tremendous wall of black clouds, with the top extending to 55,000 feet and higher. At other times, it is so weak and inconsequential

that it is very difficult to locate with certainty on a weather map, for there is little weather or cloudiness associated with it. The width of the ITCZ varies from about 20 to 150 nautical miles, and as a general rule, the narrower the zone (i.e., the greater the convergence), the more intense is the weather associated with it.

When the ITCZ is located near the equator, only small and weak cyclonic circulations (vortices) can develop within it. But when it migrates away from the equator (at least five degrees or more), the influence of the earth's rotation becomes great enough to transfer sufficient "spin" to the converging air currents to permit tropical cyclones, hurricanes, and typhoons to develop. This is only one reason why aircraft reconnaissance and weather satellites are so very important over tropical oceanic areas where weather data are normally very, very sparse.

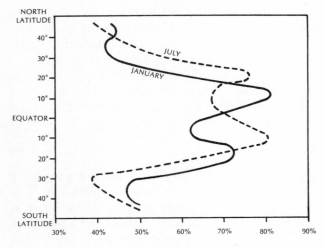

Figure 8–2. Constancy of the surface-wind direction versus latitude expressed in percent. Note that the trade winds attain a constancy of 80 percent.

EASTERLY WAVES (EWs)

In general, weathermen and other scientists talk about "waves" in a flow current when that current exhibits fairly sinusoidal oscillations. As mentioned previously, and as shown in figure 8–2, one of the most permanent air currents flowing over the earth is that of the northeast and southeast trade winds over the tropical and subtropical oceanic areas of our planet. At times, these easterly trade currents oscillate in a wave-like manner, as shown in figure 8–3, and thus weathermen speak of "waves" in the tropical easterlies, or easterly waves (EWs).

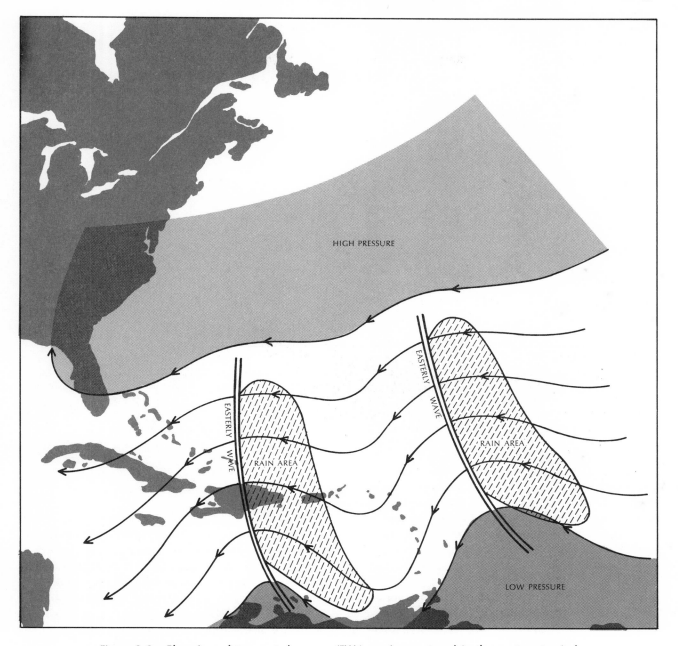

Figure 8–3. Plan view of two easterly waves (EWs) moving westward in the western tropical Atlantic region. The double lines are the EWs. Streamlines show the air flow in the lower levels of the atmosphere.

Easterly waves are extremely important phenomena because of their relation to tropical cyclone, hurricane, and typhoon formation. Basically, these waves are troughs of low pressure which are embedded in the deep easterly currents located on the equator side of the large oceanic high-pressure cells centered near 30 to 35 degrees latitude. On the average, easterly waves occur about every 15 degrees of longitude during the summer season primarily and have an average length of about 15 to 18 degrees of latitude. They extend vertically in the atmosphere from the earth's surface to, roughly, 26,000 feet and travel *from east to west* at an average speed of 10 to 13 knots. Note in figure 8–3 that easterly waves are at right angles to the air flow and that the wave amplitude decreases with increasing latitude. Rather than being distinct air-mass boundaries (such as fronts), easterly waves are zones of transition 30 to 100 miles wide, in which the weather changes gradually, but very definitely.

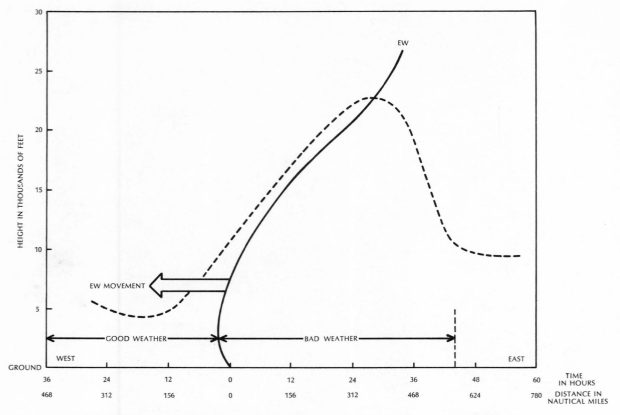

Figure 8–4. Vertical model of a standard easterly wave moving from east to west at 13 knots.
The dashed line indicates the depth of the moist layer.

Figure 8–4 is a vertical model of the normal type of easterly wave. Note that the average slope is in a ratio of 1/70 and that the EW slopes upward from west to east. This, coupled with the fact that convergence of air flow occurs to the east of the EW and divergence to the west of the EW, results in the bad weather occurring behind—to the east of—the EW as shown in figures 8–3 and 8–4.

As is true of almost all tropical disturbances, the associated cloud patterns and arrays are best displayed over the open ocean, free from the influences of land surfaces. This is still another reason why aircraft reconnaissance and weather satellites are so very important. Figure 8–5 is a plan-view sketch of the cloud patterns of a normal easterly wave as would be seen from a high-flying aircraft or a weather satellite. Note that the clouds are arranged almost in rows, forming "cloud streets." To the west of the EW, the rows of clouds are oriented from the northeast to the southwest. The clouds here are small, fair-weather cumulus clouds, dying out further west. To the east of the EW, the clouds are oriented from the southeast to the northwest. Heavy cloudiness and bad weather prevail to the east of the EW. A brief summary of easterly wave weather is contained in table 8–1.

MONSOON DEPRESSIONS

Relatively weak cyclones travel over many portions of the oceanic and continental tropics. Surface temperatures are almost constant through these cyclones, or vortices, except for a slight decrease in the area of most intense rain, where evaporation from the falling raindrops cools the atmosphere. Pressure at the center may be two to four millibars lower than in the periphery. Wind speeds average 10 to 20 knots but may increase to 30 to 35 knots close to the center on the north side. As a rule, these cyclones move toward the west or west northwest at 10 to 12 knots.

These vortices are most prominent over southern Asia during the height of the summer monsoon season, when the ITCZ is located on the Asiatic continent, as shown in figure 8–1. These cyclones, or depressions, as they are called in India, move westward, steered by a strong easterly current above 25,000 to 30,000 feet. Except for strong topographic effects, much of the precipitation in Southeast Asia is derived from these cyclones.

Table 8–1   Summary of Easterly Wave Weather

	West of Trough	*Close to Trough*	*At Trough Line*	*East of Trough*
CLOUDS	Fair-weather cumulus. Few build-ups.	Cumulus build-ups. Some high and middle clouds.	Heavy cumulus build-ups. Broken to overcast high and middle clouds.	Heavy cumulus and thunderheads. Layers of high and middle clouds.
VISIBILITY	Strong haze.	Improving.	Good, except in showers.	Fair to poor in showers.
PRECIPITATION	Usually none.	Scattered showers.	Frequent showers.	Heavy showers, thundershowers, and rain.
SURFACE WINDS	ENE to NE.	ENE to E.	Easterly, some gusts.	ESE to SSE, with frequent gusts.

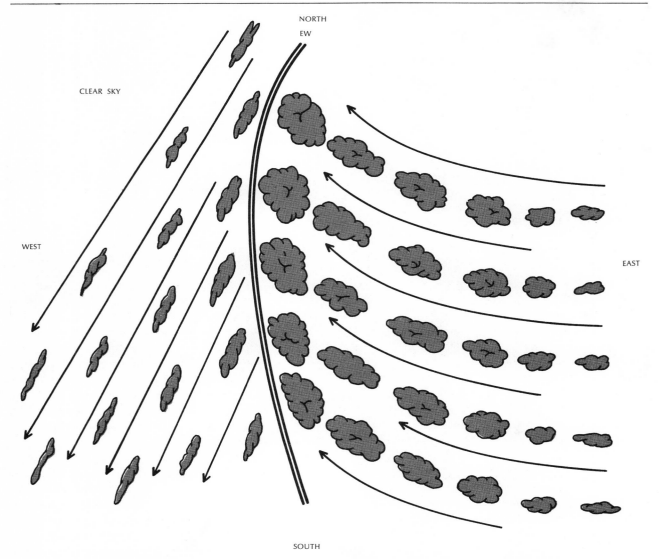

Figure 8–5.   Plan view of "cloud streets" associated with a normal easterly wave, as they might be seen from a high-flying reconnaissance aircraft (adapted from J. S. Malkus and H. Riehl)

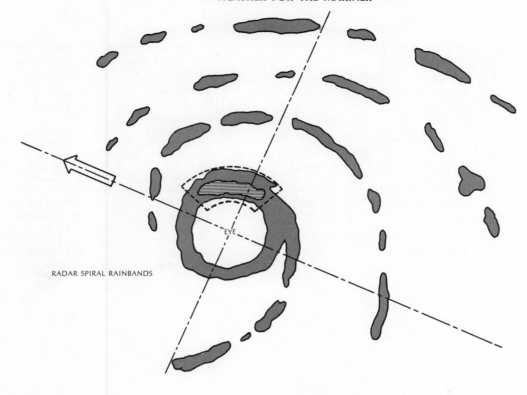

RADAR SPIRAL RAINBANDS

EYE

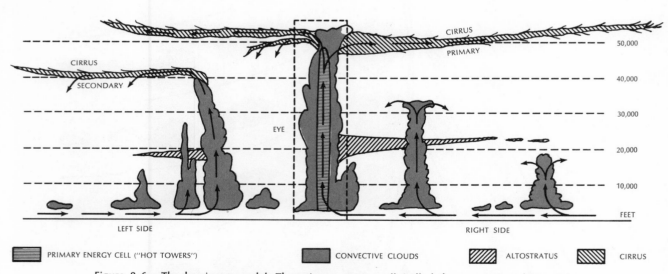

| | PRIMARY ENERGY CELL ("HOT TOWERS") | | CONVECTIVE CLOUDS | | ALTOSTRATUS | | CIRRUS |

Figure 8–6. The hurricane model. The primary energy cell (called the *convective chimney*) is located within the area enclosed by the broken line. (From Project *Stormfury* Annual Report, Appendix D, 1965.)

HURRICANES AND TYPHOONS

From a vantage point in space, hurricanes and typhoons appear as rather small, flat spirals drifting benignly on the sea—gentle eddies in the endless flowing of our planet's atmosphere. Nothing could be more misleading. Where the drift of hurricanes/typhoons (*Hs/Ts*) takes them across shipping lanes and islands and the coasts of continents, their passage is commemorated by the vast destruction of property, the great diminution of prospects—and death.

Hs/Ts are the juvenile delinquents of the tropics, the offspring of ocean and atmosphere, powered by moisture from the sea and the heat of condensation, and driven by the easterly trade winds, the temperate westerly winds, and their own violent energy. In their cloudy and spiraling tentacles and around their tranquil core, or *eye,* the winds blow with lethal velocity, the ocean develops an annihilative surge, and, as the center moves toward land, tornadoes frequently descend from the advancing wall of thunderclouds.

Compared to the great cyclonic storm systems of middle and high latitudes, Hs/Ts are of rather moderate size, and their worst winds do not attain tornado velocities. Still, their broad spiral base dominates the weather over many thousands of square miles, and they sometimes extend vertically from the earth's surface to over 50,000 feet. See the hurricane model in figure 8–6. H/T winds sometimes exceed 200 knots, and their life-span is measured in days and weeks, not hours. No other atmospheric disturbance combines duration, size, and violence more destructively than Hs/Ts. And over the centuries, seafaring men have watched the sky and ocean with anxiety for signs of an approaching hurricane or typhoon. They feared—and rightly so—encountering the small central area of these tropical cyclones with its phenomenally severe weather. Under the stress of H/T winds, mountainous waves sometimes greater than 50 feet in height are generated in the ocean. Sometimes along concavely curving coasts such as that of the Gulf of Mexico, and in estuaries, the storm surge—the rise of the water above the tide otherwise expected—often exceeds 12 feet, threatening coastal protection works and inhabited areas.

Description

As mentioned briefly in chapter 1, Hs/Ts are tropical cyclones whose associated wind speeds exceed 64 knots (75 mph). Like all cyclones, they are lows —atmospheric pressure systems in which the barometric pressure decreases progressively to a minimum value at the center, and toward which the winds blow spirally inward in a counterclockwise direction in the northern hemisphere. In the southern hemisphere, the winds spiral inward in a clockwise direction.

In some ways, tropical cyclones are similar to extra-tropical cyclones that exist as low-pressure cells along fronts, as discussed in chapters 5 and 7. The main differences are that tropical cyclones are not accompanied by fronts as extra-tropical lows usually are, and that tropical cyclones have a warm core, or center (compared to the periphery),

whereas extra-tropical lows have a cold core. Far more intense than extra-tropical cyclones, Hs/Ts often have sustained winds of 120 to 150 knots. Extra-tropical lows almost never do. Hs/Ts are about one-third as large, averaging 400 to 500 miles in diameter. They move rather slowly when heading westward, traveling in a tropical air mass of uniform temperature, rather than between air masses of different temperatures. Hs/Ts are amazingly symmetrical, with the central isobars forming almost perfect circles. They develop over only the tropical oceans characterized by extremely warm and moist air masses. They break up quickly after moving over land or over an area of very cold water, or when there is an injection of cold air into the center. Perhaps the most interesting difference between Hs/Ts and extra-tropical cyclones is the almost-calm eye found in the center of Hs/Ts, which will be discussed later in this chapter.

Formation

Many scientific papers have been published on the subject of H/T formation, but there is still no universally accepted theory. There are the convectional hypothesis, the waves in a baroclinic easterly current theory, the dynamic instability theory, and several others. However, each theory deals with only one or two specific aspects of tropical cyclone formation, and there is no composite treatment of all aspects. But this is not too important for our purposes.

The birth of Hs/Ts takes place over the tropical oceans where a condition of contrasting winds exists. This condition is found along the ITCZ, where the northeast trades (of the northern hemisphere) and the southeast trades (of the southern hemisphere) converge. Frequently, an easterly wave is present. But tropical cyclones do not develop along the equator, itself. This is because the cyclones require the "twisting force"—called *vorticity* by weathermen—of the earth's rotation to start them spinning (as discussed in chapters 2 and 5), and there is none at the equator. As seen in figure 8–1, as the ITCZ moves north of the equator in the northern hemisphere summer, the winds are twisted to the right and the cyclonic swirls are established. Thus begins the season of hurricanes in the Atlantic which batter the United States.

Most meteorologists agree on the following list of conditions as being favorable (or necessary) for hurricane and typhoon formation:

1. Surface sea temperature higher than 26°C (78.8°F).

2. Below-normal pressure in low latitudes (less than 1004 millibars, or 29.65 inches) and above-normal pressure in higher latitudes.

3. An existing tropical disturbance of some sort at the earth's surface.

4. Movement of the disturbance at a speed less than 13 knots.

5. Easterly winds decreasing in speed with height, but extending upward to at least 30,000 feet.

6. Special dynamic conditions in the air-flow near 40,000 feet.

7. Heavy rain or rainshowers in the area.

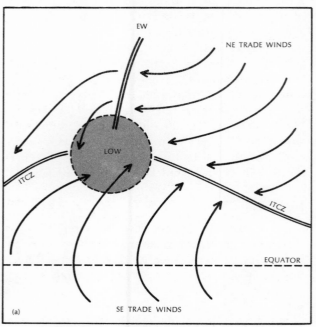

OPPOSING TRADE WINDS AND EW CAUSE AIR TO WHIRL

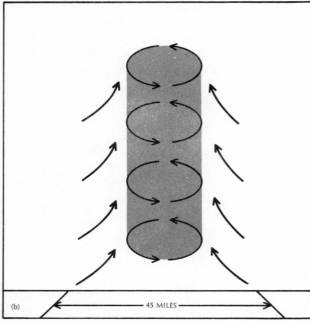

HEAT OF CONDENSATION WARMS AIR, WHICH RISES

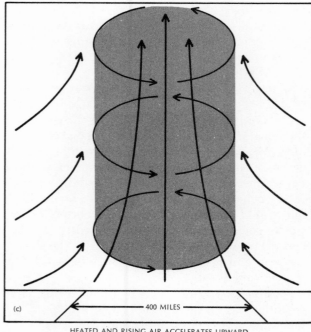

HEATED AND RISING AIR ACCELERATES UPWARD

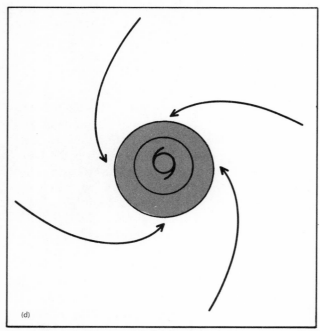

VERTICAL COLUMN IS FED FROM ALL SIDES

Figure 8–7.   The birth of a hurricane or typhoon

As mentioned previously, Hs/Ts are always born in the warm and moist air masses of the tropical oceans. In a very general way, as illustrated in figure 8–7, this frequently happens when the cyclonic motion is started by the presence of an easterly wave, followed by the intensification of the opposing trade winds whirling around each other, at least five degrees or more away from the equator. The developing and inward-spiraling winds push air toward the center, forcing the hot, moist air there to rise. This lifting process causes the water vapor to condense. The heat released thereby to the atmosphere, as the water vapor condenses, further warms the rotating air, which becomes even lighter and rises much more rapidly. As more moist tropical air spirals in to replace the rapidly rising air, more condensation takes place and more heat is released to the atmosphere. As a result, the air inside the vortex rises faster.

The vortex becomes extremely violent because of the tremendous energy released by the continuous condensation. At first, the vortex is like the inside of a giant thunderstorm. But unlike a thunderstorm over land, a hurricane has an inexhaustible source of moisture. The heat released by condensation causes the air in the hurricane or typhoon to rise faster and faster. The surrounding air rushes in with ever-increasing velocity until the hurricane is a tremendous "wheel" of inward-spiraling and violent winds.

Source Regions and Tracks

Severe tropical storms occur in all of the tropical oceans except the South Atlantic. (See figure 8–8.) This exception results from the fact that the ITCZ never migrates south of the equator in the Atlantic Ocean. The southwestern portion of the North Pacific has more tropical cyclones (*typhoons*) than any other place on earth. These are born between the Marshall Islands and the Philippines and travel first in a westward direction toward the coast of China, then northward and northeast over the Philippines toward Korea and Japan.

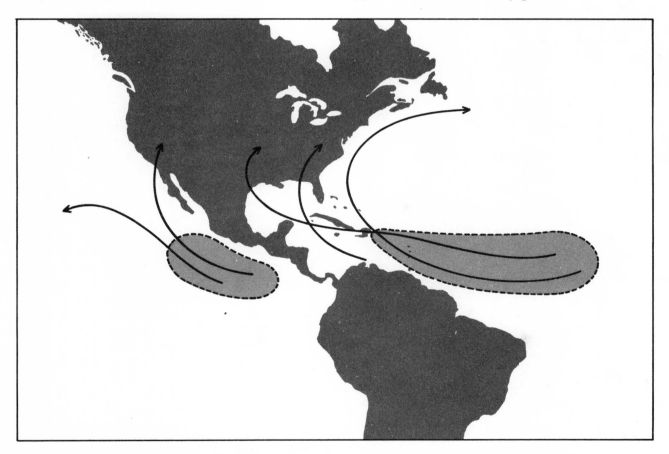

Figure 8–8. Source regions and tracks of hurricanes and typhoons in the western hemisphere. Hurricanes and typhoons generally form in the two shaded areas and travel in the direction of the arrows.

Second in frequency are the hurricanes of the South Indian Ocean (called simply *cyclones* there). Some develop north of Australia (called *Willy-Willies* there) and curve down over the northwest coast. But more frequent and intense are those that sweep westward, some brushing southeast Africa and the island of Madagascar.

Ranking third are the hurricanes that travel over the West Indies. These vortices, most of which originate near the bulging west coast of Africa, cause great damage to shipping and to the many islands in their path. These are the storms which strike the south and east coasts of the United States, or strike Central America or Mexico.

Forecasting the future tracks and movements of hurricanes and typhoons is one of the most challenging and difficult tasks that any operational meteorologist has to face. There is so much at stake.

As the result of aircraft reconnaissance, land-station and airborne radar reports, weather satellites, and the accumulation of tremendous quantities of weather data through the years, we know that simple solutions to H/T movement seldom provide the correct answer. So-called *unusual* tracks occur far too often. Too many tracks exhibit humps, loops, staggering motion, abrupt course and/or speed changes, etc. No two recorded tracks have ever been exactly the same.

In the language of most weathermen, the term *recurvature* in tropical cyclone work refers to the change in direction of movement from a westerly to an easterly component. That is, a tropical cyclone "curves" from a course of 270 degrees to 340 degrees, but it "recurves" from a course of 340 degrees to 030 degrees. Recurvature is one of the biggest problems in tropical-cyclone forecasting.

Prior to recurvature, the small-scale track of these vortices is usually close to sinusoidal, as shown in figure 8–9, leading many forecasters astray in their predictions. The speed of forward movement at this stage is normally quite slow (4 to 13 knots). After recurvature, into higher latitudes, the Hs/Ts movements are usually fairly straightline, and forward speeds of 40 to 50 knots are not uncommon.

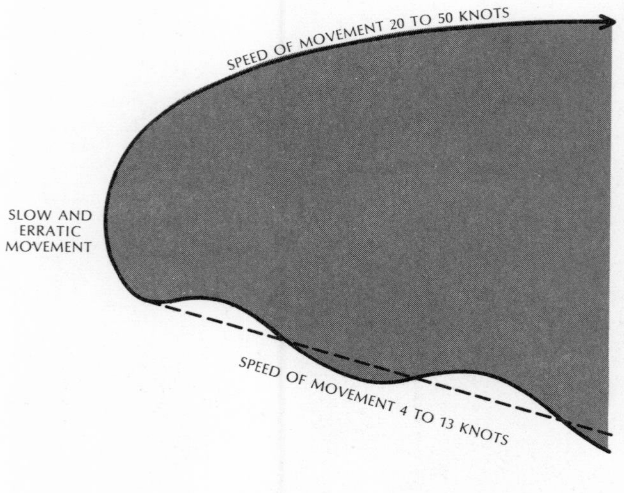

The Eye

The *eye* of a hurricane or typhoon, as seen in figure 8–10, is quite unique; it is not observed in any other weather phenomena. It has always been one of the favorite subjects of writers and investigators, for in some respects, it is the most spectacular part of the hurricane's anatomy.

As we discussed earlier in hurricanes and typhoons, the winds spiral violently inward toward the center of lowest pressure. There, the strongly converging air is whirled upward by convection, by the mechanical thrusting of other converging air, and by the pumping action of high-altitude circulations. This spiral is marked by the thick wall clouds which surround—and form the outer edge of—the hurricane's center. The wall cloud of a H/T is the area of most violent winds, heaviest precipitation, and greatest release of heat energy. This ring of strongest winds and torrential rain is usually 5 to 30 miles from the H/T's center, with the average eye diameter being about 15 nautical miles. Inside the eye wall cloud, the winds decrease rather abruptly to 12 knots or less and the torrential rain ceases. This is the celebrated H/T *eye*—a term coined many decades before the advent of radar or weather satellite pictures. The eye offers a brief, *but very deceptive*, respite from the extreme weather conditions of the eye wall. Inside the eye, the winds are light, and there are intermittent bursts of blue sky and sunlight through the thin middle and high clouds. Occasionally, low clouds form miniature spirals inside the eye. Across the eye, at the opposite wall, however, the torrential rain and violent winds resume—but come from the *opposite* direction because of the cyclonic circulation of the storm.

NORTH

SPEED OF MOVEMENT 20 TO 50 KNOTS

SLOW AND ERRATIC MOVEMENT

SPEED OF MOVEMENT 4 TO 13 KNOTS

SOUTH

Figure 8–9. Recurvature of a hurricane or typhoon

To primitive man, this fantastic transformation from storm violence to comparative calm, and from calm into violence again from another quarter—all in a relatively short interval of time—must have seemed an excessive whim of the gods. Still, it *is* spectacular. The eye's relatively calm and abrupt existence in the midst of such meteorological violence is not soon forgotten by anyone experiencing this phenomenon.

Although the eye of a hurricane or typhoon is usually described as circular, it is frequently observed by aircraft reconnaissance crews to be elliptical. When this is the case, the longest axis is usually parallel to the direction of the storm's movement. Aircraft crews have also observed that the diameter of the eye at 10,000 feet is just about twice the diameter at the earth's surface, and that the eye is constantly undergoing transformation. This is hardly surprising, however, when one considers the extreme conditions at the boundary of the eye.

Figures 8–11 to 8–15 show a superb series of radar photos taken of the scope of a Sugar Mike 10-centimeter radar. These photos are a dramatic description of the U.S. Third Fleet's devastating encounter with the eye of a typhoon in the western North Pacific in 1944. During this typhoon three destroyers capsized and went down, nine other ships sustained serious damage, and 19 other vessels received lesser damage. The magnitude of the disaster concerned Admiral C. W. Nimitz, USN (then Commander in Chief, U.S. Pacific Fleet) who wrote in part: "The time for taking all measures for a ship's safety is while still able to do so. Nothing is more dangerous than for a seaman to be grudging in taking precautions lest they turn out to have been unnecessary. Safety at sea for a thousand years has depended on exactly the opposite philosophy."

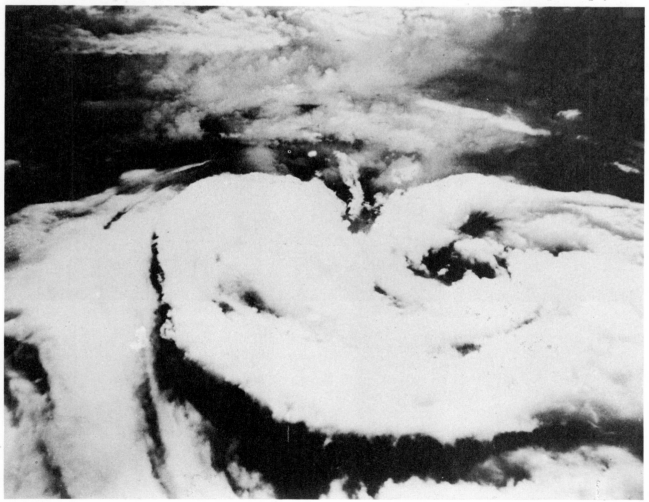

Official U.S. Navy Photograph

Figure 8–10. High-altitude view from an aircraft showing a hurricane's eye, wall clouds, and spiral rain bands

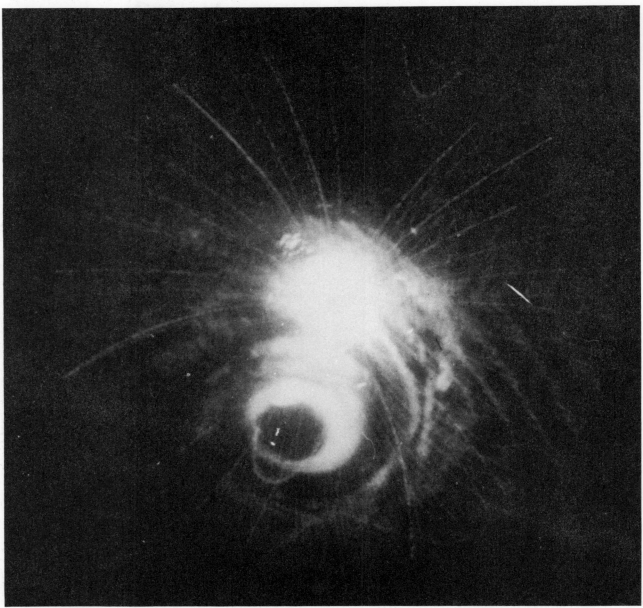

Courtesy of Captain Edwin F. Barker, Jr., USN

Figure 8–11. A radarscope's view of a passing typhoon (Situation 1)

Time	1100 Local Time
True Bearing and Distance of Eye . .	077°; 39 Nautical Miles
Wind	336°; 57 Knots; Gusts to 66 Knots
Pressure	994.5 mbs (29.37 inches)
Ceiling	Less than 500 Feet
Visibility	800 to 1,200 Yards
Waves/Swell	Very High (to 40 Feet)

Courtesy of Captain Edwin F. Barker, Jr., USN

Figure 8-12.   A radarscope's view of a passing typhoon (Situation 2)

Time	1130 Local Time
True Bearing and Distance of Eye . .	068°; 38 Nautical Miles
Wind	323°; 68 Knots; Gusts Over 80 Knots
Pressure	995.0 mbs (29.38 inches)
Ceiling	Less than 500 Feet
Visibility	800 to 1,200 Yards
Waves/Swell	Very High (20 to 40 Feet)

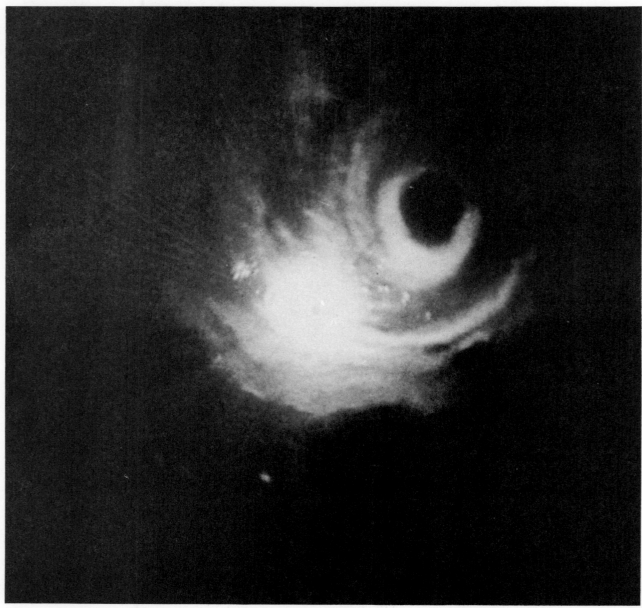

Courtesy of Captain Edwin F. Barker, Jr., USN

Figure 8–13.   A radarscope's view of a passing typhoon (Situation 3)

Time	1200 Local Time
True Bearing and Distance of Eye . .	055°; 40 Nautical Miles
Wind	315°; 65 Knots; Gusts Over 75 Knots
Pressure	996.1 mbs (29.42 inches)
Ceiling	Less than 500 Feet
Visibility	600 to 1,000 Yards
Waves/Swell	Very High to Mountainous (in excess of 40 feet)

Courtesy of Captain Edwin F. Barker, Jr., USN

Figure 8–14.   A radarscope's view of a passing typhoon (Situation 4)

Time	1340 Local Time
True Bearing and Distance . . .	005°; 38 Nautical Miles
Wind	270°; 49 Knots; Gusts Over 75 Knots
Pressure	993.2 mbs (29.33 inches)
Ceiling	Less than 500 Feet
Visibility	400 to 800 Yards
Waves/Swell	Very High to Mountainous (in excess of 40 feet)

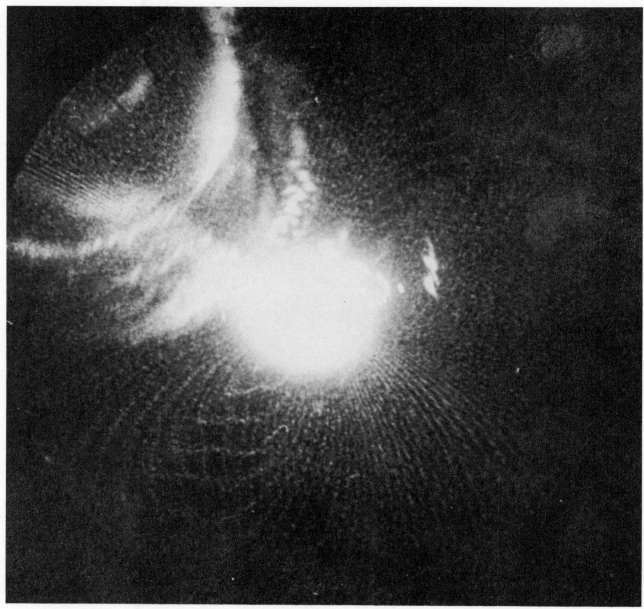

Courtesy of Captain Edwin F. Barker, Jr., USN

Figure 8–15.   A radarscope's view of a passing typhoon (Situation 5)

Time	1530 Local Time
True Bearing and Distance of Eye . .	326°; 81 Nautical Miles
Wind	250°; 41 Knots; Gusts Over 60 Knots
Pressure	1000.1 mbs (29.53 inches)
Ceiling	7,000 to 8,000 Feet
Visibility	6,000 Yards Off Bow; Less than
Waves/Swell	1,000 Yards Off Stern
	High to Very High (12 to 20 feet)

Pressure and Wind

As we discussed in chapter 5, motion in the earth's atmosphere—the wind—is a function of the pressure difference between two points. In general, the winds tend to blow across the isobars at a certain angle, from higher toward lower pressure. The greater the pressure difference, the faster the wind blows. Obviously, then, the violent winds, such as those found near the centers of Hs/Ts, are indicative of extremely large pressure differences. The pressure at the centers of Hs/Ts is very low in order that the rapid decrease in pressure in a short horizontal distance be established and maintained to support the devastating winds. In a general way, the pressure in the center of a H/T is a measure of its intensity; the lower the pressure in the center, the more intense the H/T. To the writer's knowledge, the lowest scientifically accepted sea-level barometer reading occurred in a typhoon 460 miles east of Luzon—886.56 millibars (26.18 inches).

Figure 8–16 is a typical example of a barograph trace as a H/T approaches, passes over, and leaves a ship's position or station. On the day before the H/T's arrival, local weather is frequently unusually good with the barometer reading above normal. Then, the pressure begins to fall slowly (in excess of the diurnal value) and the wind may start blowing from an unusual direction. Finally, the pressure falls at an extremely rapid rate (usually for about three hours) as the eye, itself, approaches and rises almost equally rapidly as the eye moves away. Pressure falls of as much as 40 millibars (1.181 inches) in 20 minutes have been recorded, and total pressure drops of 60 millibars (1.772 inches) in 50 miles horizontally are not uncommon.

The highest H/T winds have never been scientifically recorded simply because these instruments have collapsed and have been carried away long before the greatest intensity of the storm is reached.

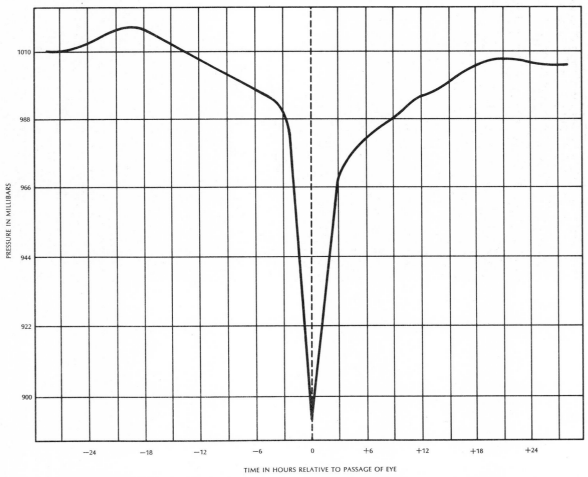

Figure 8–16. Typical example of a barograph (pressure) trace as the eye of a hurricane or typhoon approaches and passes over a ship or station (From *Heavy Weather Guide,* U.S. Naval Institute, 1965.)

Maximum estimates are based primarily on (1) aircraft reconnaissance flights, (2) calculations from pressure-gradient measurements and pressure-temperature relationships, and (3) detailed studies of structural damage. The horizontal distribution of wind speeds around a typical H/T is illustrated in figure 8–17. This diagram shows a maximum of 150 knots (173 mph) to the right of the direction of movement of the center, looking downstream. It is characteristic of a moving H/T that the strongest winds occur to the right of the center. In the case of stationary centers, the distribution of wind speeds is much more symmetrical. Figure 8–18 illustrates

the vertical distribution of wind speeds around a H/T center.

Evidence is available from gust recordings which indicates that the extreme gusts frequently exceed the maximum sustained winds by as much as 30 to 50 percent. This means that for a 150-knot H/T, one should expect extreme gusts as high as 225 knots.

One must remember that the force exerted by the wind does not increase proportionately with velocity, but rather with the square of the velocity. Doubling the wind velocity results in four times the force! Thus, a 52-knot wind results in a force of 15 pounds per square foot, but a 110-knot wind results

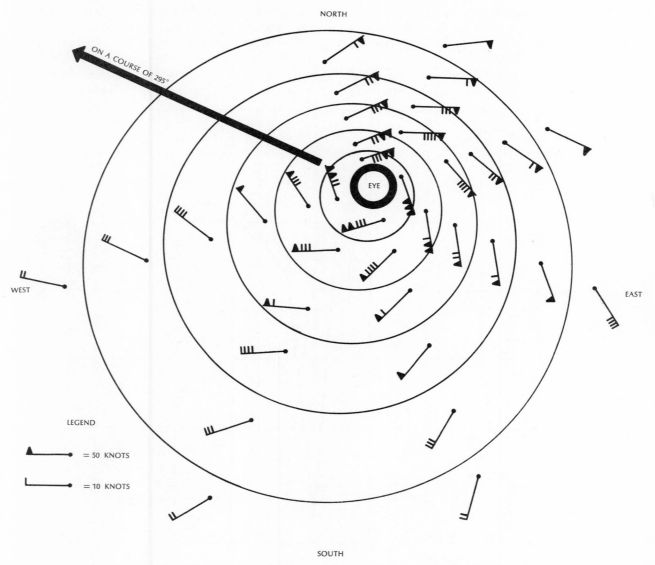

Figure 8–17. Horizontal distribution of wind speeds around a 150-knot hurricane or typhoon in the northern hemisphere. Each pennant signifies 50 knots of wind and each barb indicates 10 knots. Thus two pennants and two barbs equal 120 knots, and one pennant and two barbs equal 70 knots. (From *Heavy Weather Guide*, U.S. Naval Institute, 1965.)

in the terrific force of 78 pounds per square foot. The problem is quite complicated, however. H/T-force winds passing over and around ships and installations develop both positive and negative forces. Positive forces are those pushing in, and negative forces are those pulling out (suction effect) externally. For certain angles of incidence of the wind, the negative pressure may be considerably larger than the frontal force. The gustiness of the wind results in the uneven, intermittent pressures and wrenching effects that cause so much damage—especially to tall buildings and towers.

Clouds and Precipitation

Even though the minute details of cloud sequence varies from one H/T to another, the precursory signs are remarkably uniform. On the day before the H/T's arrival, local weather is frequently unusually good. There are only a few cumulus-type clouds, and these do not extend very high. Late in the afternoon on the day before the H/T's arrival, high-level cirrus clouds approach from the direction of the center. One begins to feel muggy. The cirrus clouds are followed in order several hours later by lowering and thickening cirrostratus, altostratus,

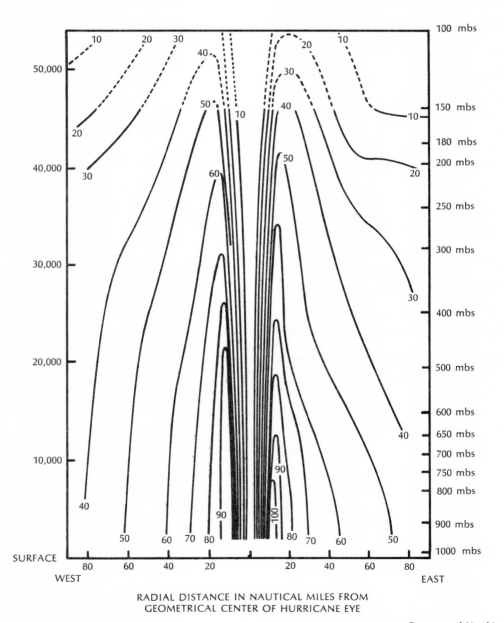

RADIAL DISTANCE IN NAUTICAL MILES FROM
GEOMETRICAL CENTER OF HURRICANE EYE

Courtesy of Hawkins and Rubsam
Figure 8–18. Vertical cross section of Hurricane Hilda (1964) showing the relative wind speeds in knots

and altocumulus clouds. Then, several brief periods of tall cumulus clouds with increasing shower activity are experienced. The showers become much heavier, more frequent, and the wind begins to increase substantially. Finally, a dark wall of clouds approaches—the *bar* of the storm. With its arrival, the full fury of the H/T is unleashed.

The full fury of the H/T is not constant. Periods of extreme violence alternate with periods of much less intensity. Torrential rain is interspaced with periods of only moderate rain. Heaviest rain occurs under the wall cloud and under the spiral cloud bands, whose shape is close to that of a logarithmic spiral. Some of the world's heaviest rainfalls have occurred in connection with Hs/Ts. In 1911, one typhoon inundated the Philippines with 88 inches of rain in a four-day period—which is almost the total annual rainfall on the island of Guam. Exact rainfall of a H/T is almost never accurately known. After the wind speed exceeds 50 knots or so, it is unlikely that the rain gauges collect more than 50 percent of the rainfall. And when the winds go higher, the rain gauges become a part of the lethal, flying debris. Total rainfall at a particular locality is, of course, largely dependent upon the speed of movement of the H/T, simply because in slowly moving storms the rain lasts much longer.

Waves, Swells, and Floods

Waves are generated by the wind, and as these wind-generated waves move out of, or ahead of, the generation area and into regions of weaker winds, they decrease in height and increase in period. These decaying waves, which persist for a long time, are termed *swells*.

The great wind stress exerted on the surface of the sea in Hs/Ts produces huge waves with phenomenal heights. Some have been reported as high as 66 feet. Figure 8–19 shows schematically how these waves travel outward in all directions from

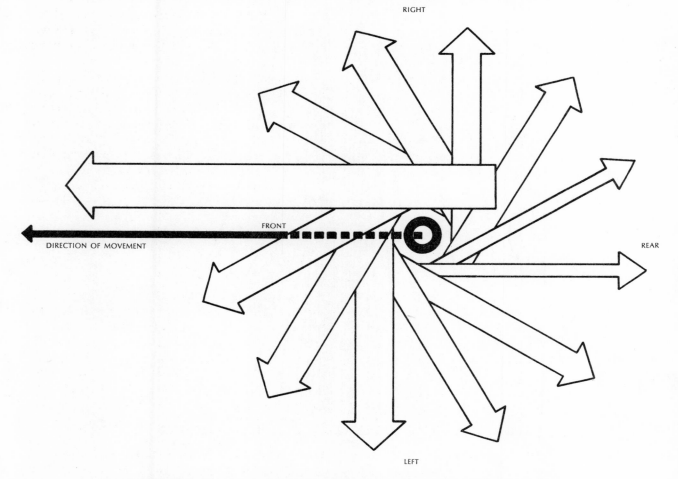

Figure 8–19. Waves/swells generated by a hurricane or typhoon. Arrows indicate the direction of movement of the waves/swells. The width of the arrows indicates relative height. (From *Heavy Weather Guide*, U.S. Naval Institute, 1965.)

the center of the H/T. These waves are observed as a long swell as much as 2,000 miles distant from the storm center. The speed of propagation of these long waves is as high as 1,000 miles per day. Since the average H/T speed of movement prior to recurvature is only 330 miles per day, these swells provide an excellent early warning of an impending H/T.

Referring again to figure 8–19, the waves generated in the right rear quadrant of the storm travel in the direction of the storm's movement. Thus, they propagate under the influence of winds having relatively little change in direction for a much longer time than the waves in any other quadrant. Consequently, the highest waves (and later, swells) are produced in this part of the storm. When these swells arrive at a distant ship (or coastline or island), the normal wave frequency of 10 to 15 per minute will have decreased to about 2 to 5 per minute. The direction of the swell indicates the position of the H/T's center at the time the swells were generated.

If the swell direction remains constant, the H/T is approaching the ship (or area) directly. If the swell changes counterclockwise, as seen by an observer facing the storm, the center of the H/T will pass from right to left. If the swell changes clockwise, the center of the H/T will pass from left to right.

One must remember that a cubic yard of water weighs three-fourths of a ton, and a breaking wave moving toward shore at speeds of more than 50 knots is one of the most destructive elements connected with a hurricane or typhoon. The erosive power of these storm-driven waves is tremendous, for these waves can scour out hundreds of feet of beach in just a few hours.

The *storm surge* occurs just to the right of the H/T's center, and either shortly precedes or accompanies the arrival of the center. When the surge moves into a long channel, however, the rise in the water at the end of the channel may reach its peak several hours after the H/T winds have passed. A H/T moving toward or crossing a coastline will always be accompanied by above-normal tides, particularly near and to the right of the center. These high tides frequently occur at a considerable distance ahead of the advancing center, providing another excellent warning sign.

Inundation by torrential rains, storm surges, and floods has been responsible for the deaths of millions of people and property damage in the billions of dollars through the years. No area affected by hurricanes and typhoons has been immune. But the extent of loss of life and destruction of property has been largely dependent upon the physical characteristics of the drainage basin, the rate and total accumulation of rainfall, and the river stages at the time the rains begin.

Rules for Precaution or Disengagement

The following precautionary rules are offered:

1. If the surface pressure drops to 1002 millibars (29.59 inches) in your location, start thinking about preliminary buttoning-up operations for your craft, whether or not you have received a "warning." If the surface pressure drops below 1,000 millibars (29.53 inches), start the preliminary buttoning-up. Your receipt of the warning may have been delayed for some reason.

2. If you notice an unusual long swell with a frequency of about two to five crests per minute, the chances are that there is a storm center at some distance from your position, in the general direction from which the swell is approaching.

3. If the center is anywhere near your position, keep a detailed plot of the six-hourly warnings which are broadcast.

4. Never try to outrun or pass ahead of a H/T center if there is some other course of action available to you. Chances are that you will *not* make it. Remember that the H/T-generated swells move ahead of the storm at speeds close to 50 knots. Once these swells arrive at your vessel, your speed of advance is cut down to four to five knots, and the eye could pass right over you. If you should become caught in the eye, don't let the relative calm fool you. Remember that after the calm, the wind will blow even more furiously from the opposite direction.

5. The winds are the best indication of the direction of the center from your craft. Memorize the diagram of figure 8–17. Adding 115 degrees to the direction from which your observed true wind is blowing gives the approximate bearing from your craft to the H/T center. For example, if your craft bears 090 degrees from the center, the observed true wind would be from 155 degrees. Adding 115 degrees to 155 degrees, the true bearing of the eye from your craft would be 270 degrees. For sums greater than 360 degrees, subtract 360 to obtain the bearing.

6. If circumstances preclude avoiding the H/T altogether, the proper action to take depends upon your position relative to the H/T's center and its direction of travel. If possible, avoid the dangerous

semicircle (the right semicircle, looking down-stream, in the northern hemisphere). Try to maneuver into the navigable semicircle (the left semicircle, looking downstream, in the northern hemisphere). The reverse is true in the southern hemisphere.

7. A gradually veering (changing clockwise) wind indicates that you are in the dangerous semi-circle. A backing (changing counterclockwise) wind indicates that you are in the navigable semi-circle.

8. If the wind direction remains steady with in-creasing speed and a rapidly falling barometer, you are near (or in) the path of the H/T. If the wind direction remains steady with decreasing speed and a rising pressure, you are on the H/T's path, but in a safe position behind the center.

9. If you are fortunate enough to have radar equipment, be sure to keep it "fired up." This will give you the bearing and distance of the eye—as long as the equipment holds up.

10. Do not guess. Work out your relative move-ment plot and the closest point of approach very carefully on a maneuvering board or your radar scope.

11. In general, if unavoidably caught within the circulation of a hurricane or typhoon—other things being equal—the best courses of action are as follows:

In the right or dangerous semicircle bring the wind on the starboard bow (045 degrees relative), hold course and make as much speed as possible.

In the left or navigable semicircle bring the wind on the starboard quarter (130 degrees relative), hold course and make as much speed as possible.

On the storm track—ahead of the center—bring the wind two points on the starboard quarter (158 degrees relative) hold the course, and make as much speed as possible. When well within the navigable semicircle, bring the wind to 130 degrees relative.

On the storm track—behind the center—steer the best riding course that maintains a constant or in-creasing distance between your position and that of the H/T. Remember, however, that Hs/Ts tend to curve northward and recurve eastward.

12. Hopefully, you will have safely moored your craft or evacuated it to a safe haven long before the storm's arrival. Once this is accomplished, do not return to the scene until all danger has passed. Remember that weather forecasters are not wizards or magicians. They, too, make mistakes, and al-lowances should be made for errors in forecasts.

Hurricane Modification by Man

Experiments in modification of hurricanes are a natural outgrowth of other research efforts and are based on facts learned about these storms since the National Hurricane Research Laboratory was formed in 1959. These experiments are a coopera-tive effort of the Departments of Defense (Navy) and Commerce (ESSA) and are grouped under the code name *Project Stormfury*. The Air Force and other governmental agencies also contribute sup-port. The objective of *Stormfury* is to determine to what extent beneficial modification of tropical cyclones, including hurricanes, is feasible.

Project *Stormfury* is primarily a field program. This is so because the theoretical models do not, as yet, simulate nature with sufficient accuracy to enable obtaining the answers to a number of im-portant questions in the laboratory.

Observations and analyses made by the National Hurricane Research Laboratory and its collaborators during the last ten years have shown that the es-sential processes responsible for the maintenance of the hurricane as a coherent weather system occur in limited regions of concentrated convection, located primarily in the main clouds around the eye, and in the principal spiral rainbands.

In the opinion of Dr. R. Cecil Gentry, Director of the National Hurricane Research Laboratory: "While actual modification of hurricanes may still be some years away, the rapid increase in knowl-edge in recent years lends support to those who believe that man will be able to exert at least some control on 'the greatest storm on earth.'"

IN MIDDLE AND HIGH LATITUDES

The traveling anticyclones (usually good weather) and cyclones (almost always bad weather) of middle and higher latitudes have al-ready been discussed in chapter 5. The subjects of weather fronts, frontal weather, and the life his-tories of frontal lows were treated in chapter 7. Consequently, these subjects will not be repeated here. It would be well to bear in mind, however, the average extra-tropical (or nontropical) storm tracks across the United States, as shown in figure 8–20. Also, take note of the mean monthly surface pres-sure charts and storm tracks for the North Atlantic area.

The pressure at the center of mid- and higher-latitude cyclones of average intensity ranges from about 988 to 1004 millibars. For strong continental cyclones, the range is about 968 to 988 millibars.

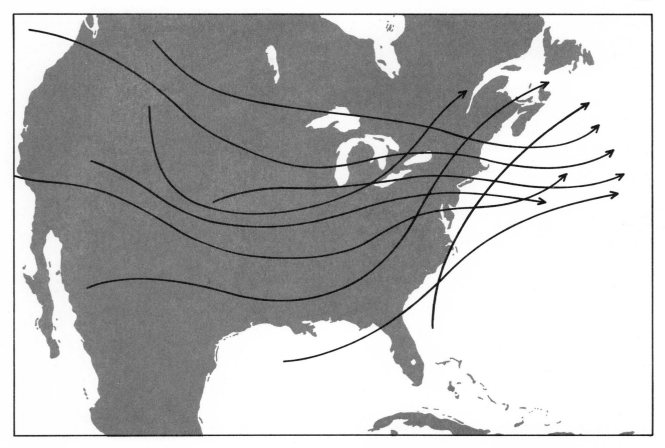

Figure 8–20. Extra-tropical storm tracks across the United States. Arrows show the paths that lows generally follow. The tracks are more widely separated in the winter than in summer.

For extreme cyclonic storms over the oceans, the range is about 920 to 968 millibars. Because pressure change is closely related to wind velocity and the occurrence of bad weather, the barometer has been rightly valued for centuries as the key instrument for the early detection of storms. A pressure drop of two millibars in three hours is a warning signal of an impending storm. A drop of five millibars in three hours is strong, and a drop of 10 millibars in three hours is extreme. Radical changes in weather, which are often abrupt, accompany the traveling cyclones of mid- and high latitudes. The intense cyclones are accompanied by high winds and, quite frequently, weather catastrophes. At sea, one should pay particular attention not only to the actual barometric reading, but also to the value of the pressure change as well.

Waves in the Westerlies and the Jet Stream

As a rule in temperate latitudes, the westerly winds of the middle and upper troposphere follow wavy paths. The distance from wave crest to wave crest or trough to trough can be long (3000 to 5000 miles) or short (1000 to 2000 miles) depending upon certain conditions. The atmospheric waves of short wavelength are the ones associated with active cyclone development and the generation of broad bad-weather areas. Figure 8–21 is a horizontal model of the average flow pattern between 15,000 and 40,000 feet. In this diagram, the distance from trough to crest is 800 miles. The horizontal mass convergence causes the counterclockwise bending of the streamlines (lines everywhere tangent to the wind velocity vectors) in the west. Horizontal mass divergence causes the clockwise bending of the streamlines forming the ridge to the east. A developing cyclone and associated surface fronts is indicated near the inflection point of the southwesterly flow. A little to the east of this point, the warm air has moved farthest northward from the tropics. (See figure 8–22.) Streamlines and isotherms (lines of constant temperature) have almost the same pattern. Often a *dome of cold air* becomes isolated from the main cold pool of air at high latitudes, as shown in figure 8–22.

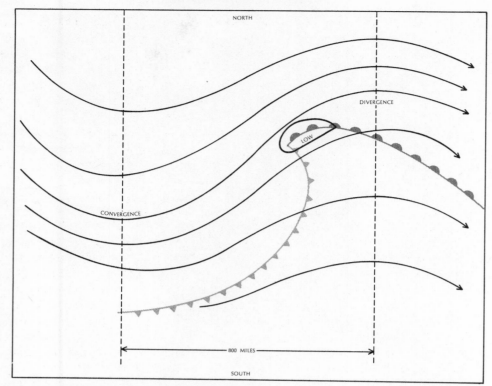

Figure 8–21.   Horizontal model of the average streamline flow pattern between 15,000 and 40,-000 feet over a developing surface cyclone. Surface cold and warm fronts are also indicated.

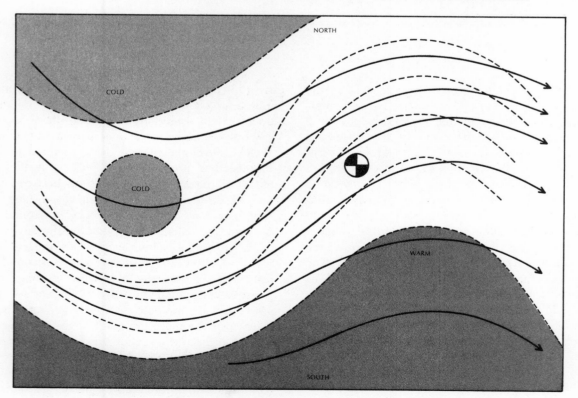

Figure 8–22.   Streamlines of the upper-level flow pattern (as in figure 8–21). The dashed lines show the isotherms during the development stage of a cyclone in the middle troposphere near 20,000 feet. The checkered circle shows the position of the cyclone center.

The tropopause—the zone that marks the top of the turbulent troposphere and the bottom of the weather-free stratosphere—was once thought to be continuous from the equator to the poles. But as the result of numerous high-altitude weather observations, we now know that the tropopause has breaks, giving it an overlapping, leaf-like structure. These breaks are important in connection with jet streams.

For our purposes, jet streams can be considered as tubular ribbons of high-speed winds in the atmosphere, usually found between 30,000 to 40,000 feet. They tend to form at the tropopause overlaps, especially at the arctic tropopause and the extra-tropical tropopause, and are at least partially the result of the strong temperature contrasts there.

On the average, jet streams are about 300 miles wide horizontally and four miles deep vertically. Wind speeds near the center—the core—sometimes reach 250 knots.

Figure 8–23 is a vertical cross-section model of a jet stream. Underneath the jet stream, the temperature decreases from right to left across the stream, looking downwind. Above the jet, the reverse is true. Thus, the jet stream is centered at the height where the temperature pattern reverses with altitude. To the left of the jet stream, the tropopause is much lower than to the right. Close to the core, a distinct tropopause often cannot be found. In this tropopause gap, the transfer of air from the stratosphere to the troposphere is thought to take place, and radioactive fallout from nuclear tests have provided the primary evidence of such a transfer. The radioactive debris often reaches the ground along a narrow belt to the right of the edge of the jet stream.

The reason that all this is so important is that the combination of short waves in the westerlies and a jet-stream core at the center of the westerly current provides the upper-air setting for the development of traveling surface cyclones and anticyclones.

The number of jet streams around and over the world and their paths vary from day to day and from season to season. Two places where jet streams occur with great frequency are over Japan and over the New England states. In winter, over the North American continent, there are usually three major jet streams: (1) over northern Canada, (2) over the United States, and (3) over the subtropics.

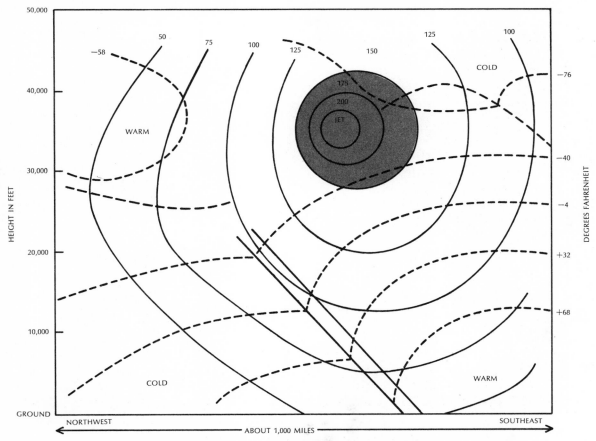

Figure 8–23. Vertical cross section of a jet stream

Thunderstorms

Thunderstorms are rather spectacular and violent local phenomena produced by cumulonimbus clouds and are characterized by squalls, turbulence, strong gustiness, heavy showers, lightning and thunder, and often hailstones which can be more than four inches in diameter. The strong, gusty winds are of serious concern to surface craft, and rightly so. Visibility is invariably poor in a thunderstorm, and ceilings are low and ragged. Thunderstorms form in a number of ways, but all require warm, unstable air of high moisture content and some sort of "lifting" or "trigger" action. The air parcels must be forced upward to a point where they are warmer than the surrounding air, so that they will continue to ascend rapidly until at some point, the air parcels have cooled to the temperature of the surrounding air. The air may be lifted in several ways—by heating, by mountains or other barriers, and by fronts.

It is quite easy to tell one's distance from a thunderstorm. We know that light travels at a speed of about 186,000 miles per second and that sound travels much more slowly, at about 1,100 feet per second, or one mile in a little less than five seconds. So the problem amounts to simply timing how long it takes for the sound of thunder to reach you after you see the lightning flash. Divide the number of seconds by five, and you have a rough approximation of the distance between you and the thunderstorm (in miles).

The development of a thunderstorm usually takes place in three fairly recognizable stages, as shown in figure 8–24. The *cumulus* stage is the first stage that each of the many cells of a thunderstorm experiences. The cumulus cloud develops into a thunderhead when the rising air currents extend to about 25,000 feet. The *mature* stage commences when the ascending air currents reach such a height that precipitation occurs. In the previous stage, the air in the cloud was warmer than the surrounding air, but now the falling rain or snow and ice crystals cool the air and downdrafts are created. Note the change in the orientation of the isotherms in figure 8–24(b), and how they dip downward in the region of the downdraft. The *final* stage occurs when the downdraft areas have increased until most of the cloud is composed of air which is sinking and therefore heated adiabatically. Since very little air is ascending and cooling, the precipitation becomes light and finally ceases. The more-or-less steady high-level winds blow the cloud of ice crystals at the top of the thunderhead into the characteristic anvil top.

In a typical summer situation, the downdraft air will arrive at the earth's surface with relatively cold temperatures of about 67°F. The drag of the falling rain also accelerates the downdraft. When it reaches the ground, the cold air spreads out on all sides, most rapidly in the direction toward which the thunderstorm is moving. At the approach of the

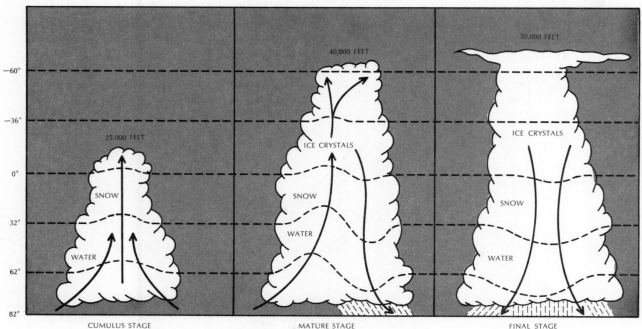

Figure 8–24. Three stages in the development of a thunderstorm. The temperature lines (isotherms) are in degrees Fahrenheit. Arrows show principal vertical motion within the cloud.

cold mass, the wind dies down, and the slowly falling barometer levels off. Then a dark line of very low scud clouds approaches rapidly. The wind shifts and begins to blow with gusts of 40 to 80 knots from the direction of the thunderstorm. The temperature falls suddenly by 20°F or more, providing some relief from the heat, and the barometer begins to rise in jerks. All this can happen in a few minutes before the main thundercloud arrives overhead, bringing with it the heavy rain. The extreme precipitation seldom lasts more than a half-hour, but in that time, one and a half to two inches of rain may fall!

Occasionally, most often in spring and early summer, thunderstorm cells form a line which can stretch as far as several hundred miles, usually oriented in a north-south or northeast-southwest direction. The line may persist for 6 to 20 hours, usually travelling toward the east. There is a vast difference between an isolated thunderstorm on a hot summer afternoon and thunderstorms along a *squall line* (as discussed in chapter 7). The latter tend to be much more severe. A *mammatus* sky, described in chapter 3, usually precedes a squall line. Incessant lightning marks the arrival of a squall line, and the wind gusts may reach close to 100 knots. Squall lines are sometimes the forerunners of cold fronts and are to be avoided by small craft if at all possible.

Tornadoes and Waterspouts

The most destructive storm in the earth's atmosphere is the tornado, a small funnel with winds up to several hundred knots revolving tightly around the core. Although the tornado is the least extensive of nature's rampages, it is, by far, the most violent and sharply defined. Tornadoes travel along paths with lengths ranging from very short distances to 70 miles or more. The width of the path of devastation may be no more than a few city blocks. On most long paths, the funnel cloud strikes the ground several times, rising after impacts a few miles long —probably a succession of new funnel formations. The funnel does little damage unless it actually touches the ground.

In the northern hemisphere, the frantic winds blow counterclockwise around the vortex. Pressure near the center of a tornado has never been recorded because no meteorological installation has been devised or built which could withstand the vortex's unbelievable fury. Meteorologists are still awaiting the occasion when a tornado will pass over a house which has a barograph in the storm cellar.

Available evidence indicates that the fantastic destruction of buildings results as much from explosion outward caused by the phenomenal decrease in pressure in the vortex, as from the violent winds. Consequently, the wisest course of action— contrary to one's natural inclinations—would be to open ports and hatches if a tornado were imminent.

Another destructive force is the 100- to 200-knot updraft at the center of the funnel. This updraft sucks houses, animals, cars, and people into the air and carries them hundreds of feet. It results from the low pressure in the vortex created by the centrifugal force of the whirling winds. It has often swept up fish and frogs from ponds, then dropped them over distant populated areas. Red clay, when so lifted, and mixed with rain later dropping to earth, has been called a "rain of blood."

Tornadoes are most common in the spring and early summer when the warm and moist air masses from the Gulf of Mexico encounter the cold air masses from the northern part of the continent. The all-time record for tornadoes in a single calendar year in the United States was broken in 1965 when a total of 927 was reported. This is almost 50 percent more than the annual mean of 628 for the 10-year period 1955 to 1964.

Again, a *mammatus* sky often gives warning of impending tornado formation. Typically, the cloud-funnel first appears near the cumulonimbus cloud base and then elongates downward. It becomes visible when condensation begins in the air that is subjected to the sudden and tremendous lowering of pressure and expansion cooling. After the funnel strikes the ground, it is blackened by soil and debris drawn into it and whirled upward.

Waterspouts are the same phenomenon as tornadoes, except that they form over the sea and are less violent. When they touch the sea surface, dense spray is drawn upward into the funnel. It has been reported that the winds in some of the less violent waterspouts rotate clockwise. This, however, has never been observed by the writer.

HURRICANE-FORCE WINDS OUTSIDE THE TROPICS

Hurricanes and *typhoons* are usually understood to be severe storms from the tropics. But for all practical purposes—and especially from a seagoing man's point of view—this definition is too limited. The threshold of hurricane- and typhoon-force winds is 64 knots (75 mph). Winds of such strength occur many times over the oceans each winter season in cyclones well outside the tropics.

As mentioned earlier, one of the main differences between tropical and extra-tropical cyclones is that the tropical cyclones have a warm core, or center, whereas the extra-tropical cyclones have a cold core. Another difference is that the extra-tropical cyclones usually have a much larger diameter. But perhaps the most important difference is the distribution of wind speeds (isotachs) around the two types of vortices.

In a tropical (warm core) cyclone, the hurricane-force winds exist in the area immediately around the center, decreasing in strength as one goes outward from the wall-cloud of the eye. But in an extra-tropical cyclone (cold core), the hurricane-force winds almost always exist in various shapes or patterns away from the center—in the storm's periphery. Figure 8–25 is a schematic diagram showing a comparison of the distribution of hurricane-force winds around the two types of cyclones. The shaded area represents the regions of hurricane-force winds (64 knots or higher).

The areas off Cape Hatteras, North Carolina, the Icelandic area of the North Atlantic, and the Aleutian area of the North Pacific, are favorite regions for the development of hurricane-force winds which generate tremendous ocean waves during the winter months. These winds sometimes blow without appreciable change in direction or speed for a distance of 1,500 miles or more, persisting for 48 to 72 hours. Over such a long distance, mountainous waves are created by the prolonged winds. Marinas, other coastal installations, homes, and other real estate are periodically exposed to the tremendous and highly dangerous force of these waves, swells, and exceptionally high tides. Who can ever forget the devastation along the entire east coast of the North American continent, extending from Canada to Florida, wrought by the storm of 6–9 March 1962?

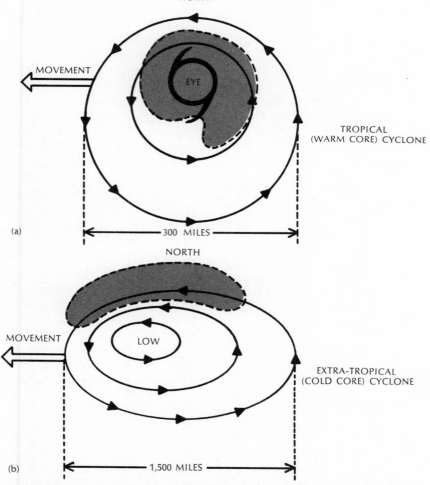

Figure 8–25. The general distribution of hurricane-force winds around tropical (warm core) and extra-tropical (cold core) cyclones. The shaded areas represent the regions of hurricane-force winds (64 knots or higher).

Chapter Nine

Forewarning and Formation of Fog

"The fog comes on little cat feet."
—CARL SANDBURG

One of the most interesting, and in some respects exceedingly important, weather phenomena is *fog* —a great swarm-like assemblage in the surface air of hundreds of thousands of water droplets so minute that it would take seven billion of them to fill a teaspoon. Expressed in the most simple terms, fog is merely a cloud whose base rests on— or very close to—the earth's surface, reducing the visibility to five-eighths of a mile or less. Although the substance of fog and cloud is exactly the same, the processes of cloud and fog formation are quite different. Clouds form chiefly because air rises, expands, and cools. Fog results from the cooling of air which remains at the earth's surface.

Fog is not only a nuisance and a menace to navigation, it has its lethal aspects, as well. During the London fog of 27 November through 1 December 1948, thousands of cars had to be abandoned by their drivers, trains and flights were cancelled, buses were led by their conductors on foot with flashlights and lanterns, and marine shipping in the Thames was at a standstill. Football matches also had to be cancelled in the zero visibility. It was estimated that this fog caused between 700 to 800 additional deaths in London. In the "killer-fog" of London from 5–9 December 1952, an estimated 4,000 people died of bronchitis and pneumonia complications as a direct result of what is now known as smog.

THE DEW POINT

Before we get into a discussion of the different types of fogs and their formation, we must say a word about *dew point* (or *dew-point temperature*). Specifically, the dew point is the temperature to which a given parcel of air must be cooled—at constant pressure and constant water vapor content —in order for saturation to occur. When this temperature is below 32°F, it is sometimes called the *frost point*. Most weathermen prefer to define the dew point as the temperature at which the saturation vapor pressure of a given parcel of air is equal to the actual vapor pressure of the contained water vapor. But why complicate things? Bear in mind the former definition and remember the following: (1) when there is a large difference between temperature and dew point and the difference does not decrease, there will be no fog; and (2) when the difference between the temperature and dew point decreases and the two values approach each other, or the difference becomes zero, look out for fog.

Let's take this point farther. The curve in figure 9–1 represents the amount of water vapor required to saturate the air at sea level for any given temperature. Note that at higher temperatures, the air —or more accurately, the space occupied by the air —can hold more water in the vapor state. The air along each point of the curve is saturated with water vapor and is at its dew point. Any further cooling will yield water as a result of condensation. Hence, fog or low-ceiling clouds will form—depending on the wind velocity. As the cooling continues, the dew point will fall because more and more vapor has condensed.

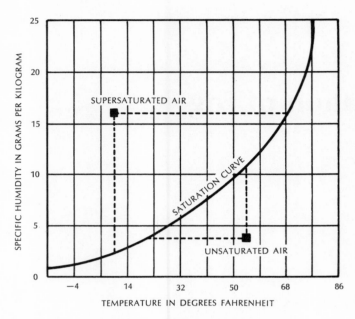

Figure 9–1.   The saturation curve

Raising the dew point is usually accomplished by the evaporation of water from falling precipitation or by the passage of a body of air over a wet surface. A high dew point is characteristic of maritime air masses. Continental air masses that travel slowly over such water surfaces as the Great Lakes show a considerable change in dew point. It can be seen from figure 9–1 that at higher temperatures a very small increase in temperature means that much more water vapor can be accommodated. A much larger change in temperature would have to be made at colder temperatures to accomplish the same increase in water content. This means that more heating by the sun is required to dispel a fog (or low-ceiling cloud) at 40°F than one at 60°F, even though the liquid content of the fog (or low-ceiling cloud) may be the same in both cases. It takes longer for a fog to dissipate in winter than in summer, partly for the reason just given and partly because there is less insolation in winter than in summer.

While stratus and cumulus cloud formations are preceded by cooling under the conservation of heat, the formation of fog involves the actual withdrawal of heat from the air, mainly through radiation cooling and the movement of air over a cold surface.

Radiation Fog

Radiation fog is a nighttime, over-land phenomenon, which occurs when the sky is clear, the wind is light, and the relative humidity is high near sunset. When the relative humidity is high, just a little cooling will lower the temperature to the dew point. Strong winds prevent nocturnal cooling. In fact, the turbulence generated by winds of only 15 knots will hinder fog formation completely. A light wind of three knots, however, is favorable for the formation of fog. The very small amount of turbulence it generates mixes the air particles which are cooled at the ground upward, thus ensuring a solid fog layer up to 40 to 100 feet. In absolutely calm air, radiation fog will be patchy and may be only waistdeep.

Because cold air drains downhill, radiation fog is thickest in valley bottoms—with the surrounding hillsides rising above the fog. Although radiation fog does not occur over water surfaces (because the heating qualities of water are such as to hold the day-to-night temperature range of water surfaces to 3°F, or less), it obscures the shore-based beacons and landmarks, and complicates navigation. Radiation fog normally disappears about one to three hours after sunrise. If it is mixed with smoke, which is especially common in winter, it forms a greasy *smog* (*sm*oke and f*og*) which may not be dissipated until late morning.

Autumn and winter are the most favorable seasons for radiation fog. During the autumn, moisture content is still high in the air, and during the winter, the nights are longest. The center of a high-pressure area is a favorite spot for radiation fog because the winds are light and the skies are usually clear (chapter 5).

Advection Fog

Advection fog may form—day *or* night—when warmer air blows over a colder land or water surface. We say the air is *advected*. This case is entirely different from radiation fog. In the case of advection fog, the warmer air gives off heat to the colder underlying surface, and this cools the air to the dew point. As the cool surface chills the warm air flowing over it, the water vapor of the air tends to condense on particles of salt, dust, or smoke, even before the relative humidity reaches 100 percent. The fog will be light, moderate, or dense, depending on the amount of water vapor that condenses.

The air in which fog forms does not tend to rise because it is being cooled, thus becoming more

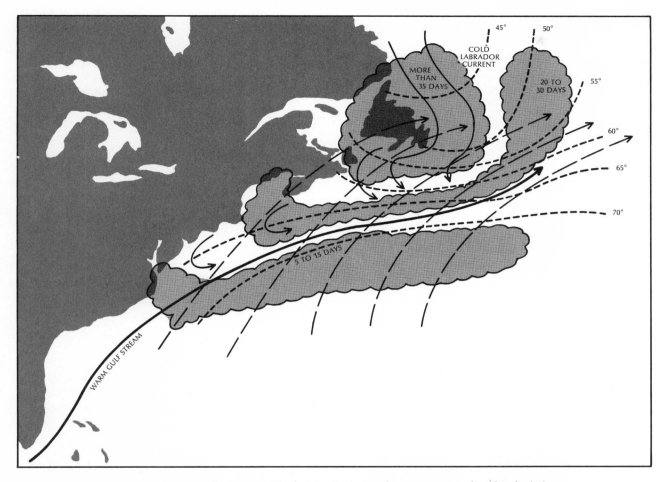

Figure 9–2. Fog over the western North Atlantic during the summer months. The shaded areas are fog; numbers in the middle refer to the average number of days with fog during the period June–August. Dashed lines are sea-surface temperatures (isotherms) in degrees Fahrenheit. Dashed arrows are streamlines of the prevailing surface air currents. Solid arrows represent the cold Labrador Current and the warm Gulf Stream.

dense, heavier, and more stable. However, if the wind speeds are 15 to 20 knots or more, over land surfaces, the air is mixed through a relatively deep layer and individual particles do not remain long enough at the earth's surface to be cooled to, or near, the dew point. Low stratus or stratocumulus clouds, rather than fog, form under these conditions. Over the sea, however, where there is less frictional turbulence than over land, advection fog will form even when the wind speeds are 30 knots. Advection fog is usually extensive and persistent.

The temperature of the earth's surface (in the northern hemisphere) normally decreases as one goes northward. Thus, advection fogs form mainly when the air currents have a direction component from the south, especially when the currents have a high moisture content.

Water areas are relatively cold for the following reasons:

1. The presence of a cold water current, such as the Labrador Current across the North Atlantic (see figure 9–2).

2. A cold upwelling of water, such as is found off the northern California coast (figure 9–3).

3. Contrasting water temperatures, such as are found along the Aleutian Islands chain where the Bering Sea is colder than the North Pacific (see figure 9–4).

4. The drainage of cold river water into a large, relatively warm saltwater area. Since fresh water is usually less dense than salt water, the less-dense cold fresh water remains on top. This sometimes establishes a cold water area over scattered parts of the northern section of the Gulf of Mexico during winter months. When water temperatures are colder than the dew point of the warm air, fog will form (see figure 9–5).

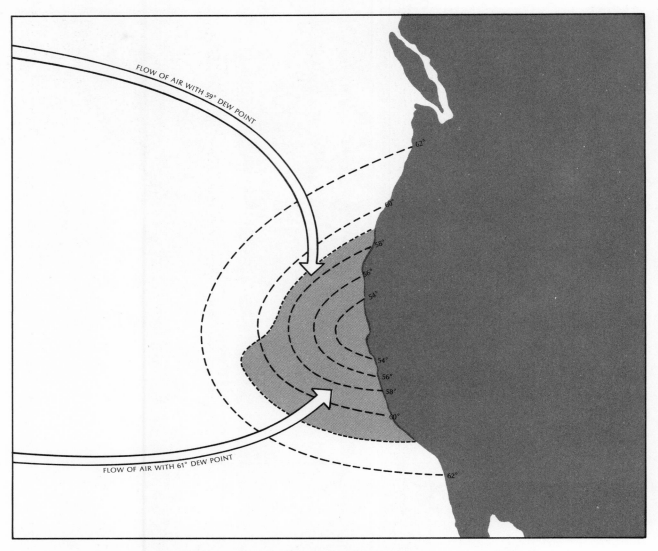

Figure 9–3. Northern California coastal fog resulting from upwelling of cold water. Shaded area is fog. Dashed lines are sea-surface temperature lines (isotherms) in degrees Fahrenheit.

Tropical-Air Fog

Tropical-air fog is really a form of advection fog which does not depend upon the flowing of warm air over a cold current, but rather on the *gradual* cooling of the air as it travels from lower to higher latitudes. It occurs over both water and land and is probably the most common type of fog over the open sea. In the United States, it forms some of the most widespread fogs observed anywhere. Tropical-air fog is more common over water than over land because of the smaller frictional effect over water. See, again, figure 9–2.

Frontal Fog

As mentioned briefly in chapter 7, fog sometimes forms ahead of warm fronts and occluded fronts and behind cold fronts. This happens when the rain falls from the warm air above the frontal surfaces into the colder air beneath. Evaporation from the warm raindrops causes the dew point of the cold air to rise until condensation on the nuclei present takes place.

Upslope Fog (Expansion Fog)

When air travels upslope for a long time over rising terrain, the decrease in temperature which it undergoes because of adiabatic expansion may lead to fog. Across the great plains of the United States, the standard atmosphere pressure decreases by about 130 millibars from near the Mississippi to the eastern edge of the Rockies. Surface air traveling this route from east to west undergoes a tem-

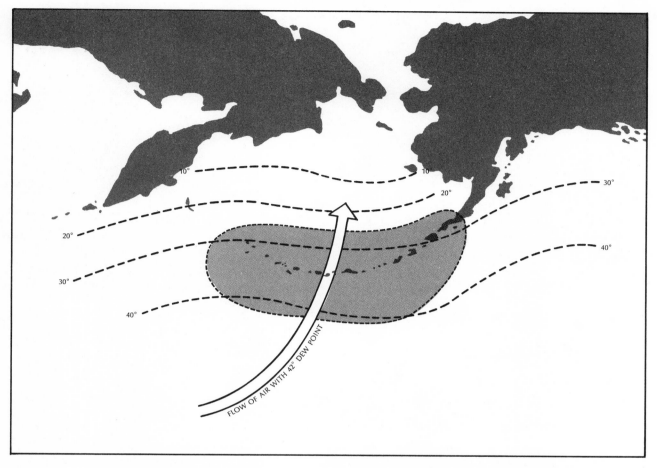

Figure 9–4. The famous Aleutian fog that persists even when the winds reach 35 knots. The shaded area is fog. Dashed lines are sea-surface temperature lines (isotherms) in degrees Fahrenheit.

perature decrease of about 23°F. This westward motion occurs mainly in late winter and spring. At that time, the difference between temperature and dew point in the lower plains is sufficiently below 23°F, so that the air traveling over the slowly rising terrain will form upslope (or expansion) fog about 100 to 200 miles to the east of the mountains.

Steam Fog and Arctic Sea Smoke

During the early morning hours in summer, long columns of steam, or mist, often rise over inland bodies of water, lakes, and even river valleys. At that time of year, water temperatures and vapor pressure over water are at their highest. Cold air draining down the slopes may be 18°F, or more, colder than the water. The water evaporating from the surface may supersaturate this air almost immediately. The evaporated water recondenses and rises with the air that is heated from below. Steam fog is usually very shallow and looks like tufts of whirling smoke coming out of the water.

When very cold air in winter moves from a land mass out over water which is perhaps 45°F warmer than the air, or when Arctic air travels from the ice shelves onto the open ocean, the steaming is so intense that it consolidates into a *fog* and is called *Arctic sea smoke.*

Inversion Fog

Inversion fog is typical of the subtropical west coasts of continents—California, for example. The upwelling of cold water is common along west coasts because of the wind action, and the air flowing over this cold water acquires a low temperature and high relative humidity. Above the cool, moist layer of air is a temperature inversion (increase of temperature with height) which acts as a "lid" and prevents the moist air from rising. See again, figure 9–3. Fog forms, similar to the advection process, and is very frequent offshore. At night, as the land cools, the fog works its way inland.

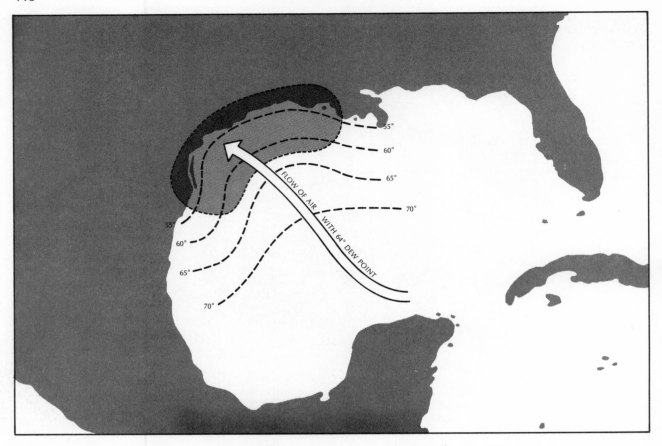

Figure 9–5.   Fog caused by the drainage of cold river water into the Gulf of Mexico during the winter months. Shaded area is fog. Dashed lines are sea-surface temperature line (isotherms) in degrees Fahrenheit.

Ice Fog

Ice fogs occur at temperatures of −22°F or colder, mostly in inhabited areas. Aircraft running up their engines in temperatures of −40°F or lower will quickly cause an airport to be enveloped with ice fog. Even animals, after a bit of exercise, find themselves surrounded with ice fog caused by the evaporation of body liquids which condense and freeze in the bitterly cold air.

Fog Over the Sea

Fog is rare at the equator and in the trade-wind belts, except along the coasts of California, Chile, and northwest and southwest Africa. But it is a common phenomenon in the middle and high latitudes, especially in spring and early summer. The Gulf Stream and the Labrador Current converge in the Newfoundland Banks region. When warm air from the Gulf Stream overruns the cold water of the Labrador Current, dense fog banks result. Similarly, in the northwest Pacific, fog is common off the coast of Asia where the warm Japanese Current overruns the cold Kamchatka Current.

During the spring and summer months, warm and moist air currents which flow from land to sea produce fog over coastal water areas. A shift in wind direction tends to drift the fog back over the adjacent land areas. During the fall and early winter months, the air currents blowing from sea to land tend to produce fog over coastal sections. Such fog may drift to sea with a reversal of the wind direction.

In the North Pacific Ocean and Bering Sea, broad areas of dense fog are common, and at times the fog extends to heights of 5,000 feet or more. It is caused by air moving northward from the high-pressure areas of the Pacific Ocean centered near latitudes 35 to 40 degrees north. The air is cooled to its dew point in passing over the colder waters to the north. This cooling of the air also increases its density, so that there is little tendency of the air to rise. See, again, figure 9–4. Fog persists in the Aleutian area even when winds are 35 knots. During the fall and spring months, fog is at a minimum in the Aleutian area. In winter, Arctic sea smoke is quite common, and at times it may build upward to several thousand feet thick.

Forewarning of Fog

Perhaps the most important single factor to remember in connection with fog is the difference between air temperature and dew point—called the dew point *spread*. Whenever the formation of fog is anticipated or predicted, it is well to keep a plot of this difference (the spread), and watch the trend closely. As the spread approaches zero, fog is imminent. Table 9–1 contains a summary of the fog-producing and fog-dissipating processes in the atmosphere. Figures 9–6 through 9–11 are classic examples of some of the different types of fog.

Table 9–1   Summary of the Fog-Producing and Fog-Dissipating Processes in the Atmosphere.

Fog-Producing Processes	*Fog-Dissipating Processes*

A. Evaporation from:
 	
   1. Rain that is warmer than the air (frontal fog and rain-area fog).
   2. Water surface that is warmer than the air (steam fog).
	A. Sublimation or condensation on:
	
	   1. Snow with an air temperature below 32 deg. F. (except ice crystal fog).
	   2. Snow with an air temperature above 32 deg. F. (melting snow).

B. Cooling from:
   1. Adiabatic upslope motion (upslope fog).
   2. Radiation from the underlying surface (radiation fog).
   3. Advection of warmer air over a colder surface (advection fog).
	B. Heating from:
	   1. Adiabatic downslope motion.
	
	   2. Radiation absorbed by the fog or by the underlying surface.
	   3. Advection of colder air over a warmer surface.

C. Mixing:
   1. Horizontal mixing (unimportant by itself and strongly counteracted by vertical mixing).
	C. Mixing:
	   1. Vertical mixing (important in the dissipation of fogs and the production of stratus clouds).

Courtesy of Hunting AeroSurveys, Ltd.

Figure 9–6.   Ground fog

Courtesy of R. K. Pilsbury

Figure 9–7.   Radiation fog in the evening. The photograph was taken after the lights were turned on, showing the patchiness and variation of the fog with height.

Courtesy of R. K. Pilsbury

Figure 9–8.   Sea fog. The fog is less dense in the warmer waters of the bay on the left, but quite thick on the right, especially where the air is forced to rise over the hills.

Courtesy of R. K. Pilsbury

Figure 9–9.   A broad belt of fog forming as relatively warm, moist air is forced to rise over the hills. Note that conditions are still quite clear at sea level, but not for long.

Courtesy of R. K. Pilsbury

Figure 9–10.   Fog forming in a valley below the level of the observer

Courtesy of R. K. Pilsbury

Figure 9–11.   Fog rising and clearing from a valley. Note the individual puffs of fog in the sun's rays to the left.

Chapter Ten

Ice Accretion—Hazard of Significance to Seafarers

The accretion of ice on ships and small craft in cold waters has not, until very recently, been given a great deal of thought by individuals other than those directly affected by it. Yet, it constitutes an especially serious hazard to the smaller ships and small craft. But even for the larger ships, including warships, the accumulation of ice on deck, on super-structure, and on equipment, greatly impairs their overall efficiency—and sometimes, their safety.

The added weight of ice reduces freeboard, and therefore reduces the range of stability of the vessel. Even more dangerous, however, is the ice formed high on masts, rigging, and superstructure. This ice has a large heeling lever, and has a most serious effect on stability. The vessel may become top-heavy and capsize. Compounded with the danger from the loss of stability is the effect of the windage, or "sail area," of the ice on masts and rigging which may make the ship or small craft difficult to handle in heavy weather. For example, it may become impossible to stay head to the wind. In addition, the accumulation of ice on aerials may render the radio and radar inoperative. It is now well known that many trawlers have been lost at sea because of ice accretion.

THE NATURE OF ICING AT SEA

Three of the basic mechanisms of ice accretion are (1) freezing rain, (2) Arctic frost smoke, and (3) freezing spray. Each of these will be discussed very briefly in the following sections.

Freezing Rain

Freezing rain will cover a ship or small craft with fresh-water glaze ice, but the accumulated weights of ice are unlikely to be sufficient to endanger the vessel directly. Rates of ice accretion can be expected to be commensurate with rates of accumulation of rainfall, and these are not great, especially in the colder climates.

Arctic Frost Smoke

As mentioned in chapter 9, Arctic sea smoke occurs when the air is at least 16°F colder than the sea. If the air temperature is below 32°F, then the Arctic sea smoke is called *Arctic frost smoke*. This frost smoke is often confined to a layer only a few feet thick, and trawlermen in northern waters refer to it as *white frost* when the top of the layer is below the observer's eye level. It is referred to as *black frost* when it extends above the observer.

The small water droplets in the frost smoke are supercooled. On contact with the vessel, part of the droplet freezes immediately, while the remainder stays liquid for a short time before it, too, freezes in the cold air. The result of the instantaneous freezing of the supercooled droplets is an accretion of opaque, white rime ice with imprisoned air. This rime ice is easier to remove than the clear ice, or glaze, which forms in other circumstances, because rime ice is porous.

Occurrences of icing on the British fisheries research vessel *Ernest Holt* have been reported in great detail. In one case, about 100 miles east of Bear Island, the ship was in air temperatures colder than 14°F, with dense frost smoke. In about 12 hours, four inches of rime ice had accumulated on deck, with a 12-inch thickness of ice on the ship's side at the level of the rail. It was calculated that the rate of accretion of ice was as much as two and a half tons per hour in this case!

Freezing Spray

Freezing spray is the most dangerous form of icing. It occurs when the air temperature is below the freezing temperature of the sea water, about 27°F. The spray freezes on the exposed surfaces of the vessel to produce clear ice or glaze. At lower air temperatures, the ice may be opaque and this may be due to the spray being supercooled so that it partially freezes on impact. At extremely low temperatures, such as 1°F and below, as might be encountered in anchorages or close inshore, wind-induced spray may be frozen before it strikes the vessel, so that it does not adhere to the vessel, but may form drifts on deck. This would not necessarily apply to any spray generated by the movement of the vessel through the water, which would strike the vessel while still liquid, and adhere. Sea ice formed from green water coming inboard will seriously add to the accumulation of ice if it is trapped on deck by ice-choked rails or freezing ports.

With air temperatures below 28°F, freezing spray is observed in winds of 18 knots or higher. The lower the air temperature and the stronger the wind, the more rapid is the accumulation of ice. A low sea temperature also increases the rate of accumulation of ice. To the spray blown from the wave-caps is added the spray generated by the vessel herself, so that the total rate of ice accretion will also depend on the design and loading of the vessel, on her heading and speed relative to the waves, and also on the relative wind (which determines which part of the vessel is most exposed). It should be understood that an accumulation of ice will, itself, increase the rate of accumulation, since the ice already formed increases the effective cross-section of rigging, mast, rails, etc., exposed to the spray. Thus, there are just too many variables for simple, precise rules to be formulated. However, figures 10–1 through 10–4 will prove a very useful guide in your forecasting the relative severity of icing conditions as determined by weather factors. These figures are based on the outstanding research in this field by Dr. H. O. Mertins.

It is interesting to note that at the investigation conducted into the loss of the trawlers *Lorella* (559 tons) and *Rodbrigo* (810 tons), the ice accretion from freezing spray on these ships amounted to 50 tons or more in 24 hours, and this was sufficient to cause them to founder in the rough seas. Weather conditions at the time were estimated to be as follows: (1) air temperature 25°F, (2) seawater temperature 34°F, and (3) wind speeds near 50 knots. As a matter of interest, plot these conditions on figures 10–3 and 10–4.

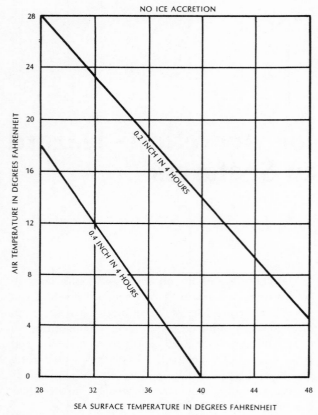

Figure 10–1. Graph for estimating rate of ice accretion when wind speeds are between 22 and 34 knots

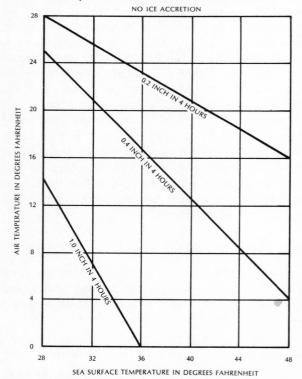

Figure 10–2. Graph for estimating rate of ice accretion when wind speeds are between 34 and 40 knots

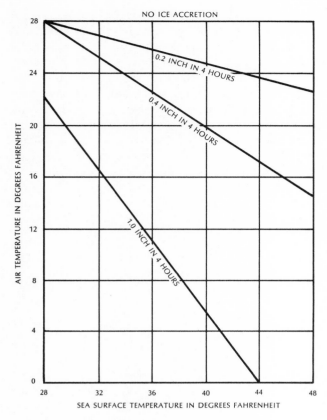

Figure 10–3. Graph for estimating rate of ice accretion when wind speeds are between 41 and 55 knots

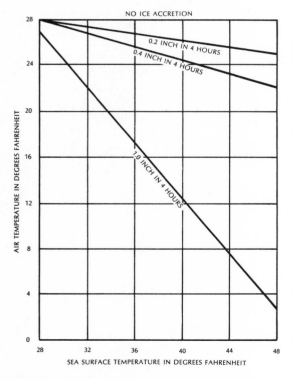

Figure 10–4. Graph for estimating rate of ice accretion when wind speeds are 56 knots or greater

No icing has been observed where the sea surface temperatures were above 41°F. There have been cases of icing with sea surface temperatures between 37° to 41°F, but the accretion of ice did not constitute a real threat.

Japanese experiments in the western North Pacific have confirmed the accretion of ice at wind speeds of 18 knots and above, with air temperatures colder than 28°F. Also, some ice accretion even occurred in wind speeds of only 13 knots. The relationship between the rate of accretion of ice and the various weather factors was not determined, but a particular accumulation measured was two tons per hour (on a 400-ton ship) in a wind of 18 knots with the air temperature 20°F and the sea temperature 32°F. In the course of these experiments, accumulations of ice up to 26 tons were frequently observed, and in one case, an accumulation of 60 tons of ice on a 500-ton ship in a period of only 24 hours was observed!

METEOROLOGICAL ASPECTS AND PREDICTIONS

Despite the many uncertainties about the quantitative aspects of ice accretion on vessels, it is evident that the worst conditions are met in the combination of very low air temperatures and strong winds. Although other circumstances also give rise to these conditions, the weather features characteristically producing them are found to the rear of a low-pressure system and on its poleward side. Since cold air will be warmed by passing over warmer water, it is clear that the coldest air will be encountered when the wind is blowing directly off a cold land area (or ice).

Because of the uncertainties concerning the rate of accretion of ice, and also because it depends, in part, on such factors as vessel design, and course and speed relative to both waves and wind, it must be realized that the wording of any predictions relating to the rate of ice accretion will be. quite difficult for the weatherman, and any scale of intensity utilized will necessarily be a coarse one.

AVOIDING THE HEAVIEST ICING

Complete protection from shipboard icing involves avoiding areas of strong winds where the air temperatures are below 28°F. When the avoidance of a strong-wind area is not practicable, then the severity of icing can be reduced by heading for less cold air and warmer water. As can be seen from figures 10–1 through 10–4, the sea surface temperature has a direct influence on the rate of ice accretion. In addition, warmer sea temperatures frequently mean warmer air temperatures. Since the air temperature also has a strong influence on the rate of ice accretion, the benefit of warmer water may be two-fold.

Vessels that seek shelter in the lee of land may still experience low air temperatures, but there should be some reduction in wind speeds and less spray blown from the wave-caps. Ship-generated spray in the calmer water will be greatly reduced.

Furthermore, attempts to remove the ice will not be thwarted by the vessel's motion and seas sweeping the deck, as would be the case in the open sea.

In very high latitudes, however, do not seek shelter in the lee of the ice edge. The ice provides negligible shelter from the wind, and here the coldest air and sea temperatures are found, and provide the most severe conditions for icing. If the wind backs or veers parallel to the ice edge, the air temperature remains very low and heavy seas are soon generated.

Chapter Eleven

Wind, Waves, and Swell

It is pleasant, when the sea is high and the winds are dashing the waves about, to watch from shore the struggles of another.
—Lucretius, 99–55 B.C.
De Rerum Natura

Mankind has been greatly interested in the oceans since before the time of Aristotle, who wrote a treatise on marine biology in the fourth century B.C. But Man's interest in the sea is not surprising when one considers that the volume of the earth's oceans is something like 328 million cubic miles, and that the volume of all land above sea level is only 1/18th of the volume of the oceans. The early studies of the ocean were concerned with problems of commerce. Information about tides, currents, and distances between ports was required.

When he was Postmaster General, Benjamin Franklin prepared temperature tables by means of which navigators could determine whether or not they were in the Gulf Stream. This resulted in faster mail service to Europe. However, the beginning of modern oceanography is usually considered to be 30 December 1872 (almost 100 years ago), when HMS *Challenger* made her first oceanographic station on a three-year round-the-world cruise. This was the first purely deep-sea oceanographic expedition ever attempted. Analysis of the sea water samples collected on this expedition proved for the first time that the various constituents of salts in sea water are virtually in the same proportion everywhere. But even before the *Challenger* expedition, Lieutenant Matthew Fontaine Maury of the U.S. Navy (often called the father of American oceanography) was analyzing the log books of sailing vessels to determine the best oceanic routes. He did much to stimulate international cooperation in oceanography and marine meteorology.

The question is frequently asked, "What is the difference between hydrography and oceanography?" Perhaps the best way to answer this is to compare the ocean to a bucket of water. Then, hydrography would be a study of the bucket, and oceanography would be a study of the water. Hydrographers are concerned primarily with the problems of navigation. They chart coastlines and bottom topography. A hydrographic survey usually includes measurement of magnetic declination and dip, tides, currents, and weather elements. Oceanography is concerned with the application of all physical and natural sciences to the sea. It includes the disciplines of physics, chemistry, geography, geology, biology, and *meteorology*. The interrelationship of specialties is one of the main characteristics of oceanography, and many times the words *hydrography* and *oceanography* are used interchangeably.

Although the subject of oceanography has many fascinating aspects, we will necessarily limit. our discussion to wind-generated waves and swell— perhaps the greatest single oceanographic hazard to navigation in coastal waters and on the open sea.

WAVES, SEA, SWELLS, AND BREAKERS

Professional mariners live in intimate contact with the waves of the sea and are able to realize better than most people the extent to which wave size and energy (as demonstrated by the destructive power of waves) are related to the speed of the wind.

They are also accustomed by their training and experience to make frequent estimates of the surface wind speed by noting the appearance of the sea surface. For the amateur mariner or weekend sailor, this is not so easy. Figures 11–1 through 11–7 show the appearance of the surface of the sea at graduated wind speeds ranging from 8 to 120 knots and constitute an excellent wind-speed/sea-state catalog with which the reader should become thoroughly familiar—especially if there is no anemometer (an instrument which indicates wind speed and direction) on board.

Official U.S. Navy Photograph

Figure 11–1.   Wind at 8 knots. Sea has occasional whitecaps.

Official U.S. Navy Photograph

Figure 11–2.   Wind at 12 knots. Occasional small whitecaps are elongated, move perpendicularly to the wind direction, and carry downwind a short distance.

Official U.S. Navy Photograph

Figure 11–3. Wind at 25 knots. Larger elongated waves begin to form. Numerous patches of white foam remain from breaking waves. Wind streaks become more noticeable.

Official U.S. Navy Photograph

Figure 11–4. Wind at 35 knots. Light spray begins to blow off the breaking waves and is carried downwind.

Figure 11–5.   Wind at 45 knots. Wind streaks become continuous.

Figure 11–6.   Wind at 60 knots. Entire sea takes on a whitish-green cast.

Official U.S. Navy Photograph
Figure 11–7. Wind at 90 knots. Sea becomes whiter and whiter from blowing spray and foam.

There is a large distinction between waves, sea, swells, and breakers. First of all, contrary to popular belief, waves are *not* masses of horizontally-travelling water particles. The wave *shape* moves horizontally, but the individual water particles—except in cases where large friction is involved—describe orbital circular motions, but remain essentially in place with little forward movement. If we throw a cork over the side, we notice that although the waves move, the surface of the water does not travel with the waves. Instead, the cork—and every other point on the water surface when waves are present—describes a circle in a vertical plane, moving upward as the wave crest approaches, forward as the crest passes, downward as the crest recedes, and backward as the trough passes. The diameter of this circle is equal to the height of the waves, and the time in which the cork moves once around it is equal to the *period* of the waves. Similarly, the movement of every point on the surface of a water wave takes place in a vertical circle. Thus, the profile of the wave surface is known as a *trochoid*.

Waves are generated in four fundamental ways in the open sea: (1) by changes in atmospheric pressure, (2) by the wind acting for long periods of time on the water's surface, (3) by seismic disturbances such as earthquakes, and (4) by the tidal attraction of the sun and moon. These methods of generation act at different times, in different directions, and in different amounts. When waves of different characteristics meet in a relatively small area, a complex pattern of wave motion results, caused by waves reinforcing or interfering with each other in varying amounts. This complex situation is termed *sea* and is very difficult to describe mathematically. *Wind waves* are defined as waves which are growing in height under the influence of the wind. *Swells* are defined as wind-generated waves which have left their area of generation and have advanced into regions of weaker winds or calms, and are decreasing in height. Swells assume a sort of regular undulating motion which closely approximates the so-called "ideal wave."

When waves enter shallow water, the friction caused by the bottom will tend to slow down the water particles nearest the bottom, while the upper particles continue with the faster original velocity. When the upper particles are traveling at a critical speed with respect to the bottom particles, the wave first steepens and then breaks. Waves in this condition are referred to as *breakers*.

Basically, the wave spectrum varies from short-period wind-generated waves with periods of 3 to 30 seconds through long-period tidal waves with periods of almost 13 hours. The term *tidal wave* refers to those waves generated by the action of the sun and moon on the oceans and should not be confused with seismic-generated waves (*tsunamis*) which have been erroneously called tidal waves.

From a mariner's point of view, the waves that most affect seamanship are those generated by the winds. It is common knowledge among seamen around the world that the longer and harder the wind blows, the higher will be the waves.

The three most important variables regarding wave relationships are the *wind* at the sea surface, the *fetch* (the stretch of water over which the wind blows), and the *duration* (the length of time that the wind has blown). Since wind, fetch, and duration vary within very wide limits, it has been necessary for meteorologists and oceanographers to develop and design special tables, charts, nomograms and slide rules to forecast all the possible variations. Although knowledge of the way in which the wind produces waves on the surface of the sea is still not complete, marine meteorologists are doing an excellent job of predicting the various oceanographic parameters. Tests made of the various methods to date indicate that wave/swell forecasts can be made with equal or greater certainty than most meteorological forecasts. The reason for this is that changes in the oceans take place much more slowly than in the atmosphere.

WAVE CHARACTERISTICS AND RELATIONSHIPS

Referring to figure 11–8, a wave is described by its *length*, *L*, (the horizontal distance from crest to crest or trough to trough), by its *height*, *H*, (the vertical distance from trough to crest), and by its *period*, *T*, (the time interval in seconds between the appearance of two consecutive crests at a given position). A wave may be *standing* or *progressive*, but our discussion will deal with progressive waves only. In a progressive wave, if the length and energy are constant, the wave height is the same at all localities and the wave crest appears to advance with a constant speed. During one wave period *T*, the wave crest advances one wave length *L*, and the *speed* of the wave is therefore defined as $C = L/T$. Waves of small height are those for which the ratio of height to length is 1/100 or less. Waves of moderate height are those for which the ratio of H/L is from 1/100 to 1/25. Waves of great height are those for which the ratio of H/L is from 1/25 to 1/7. Theoretically, the wave form becomes unstable when the ratio of H/L exceeds 1/7. Observational evidence, however, indicates that instability occurs at a steepness as small as 1/10. The wave speed increases with increasing steepness (increasing values of H/L), but the increase of speed never exceeds 12 percent. The speed of a wave (in knots) can be calculated by the following simple formula:

Velocity (knots) =
Length (feet)/Period (seconds) x 0.6.

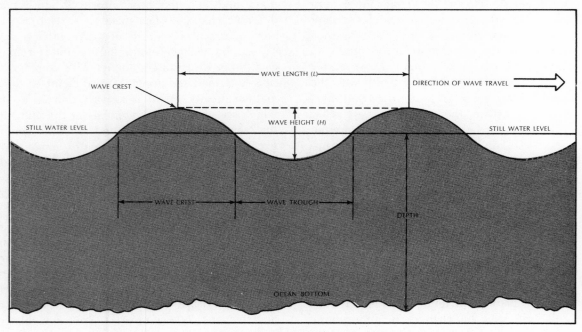

Figure 11–8. Ocean wave characteristics

Probably the greatest single item of controversy regarding waves is that of *maximum* wave height, because of the general tendency to underestimate small wave heights and to overestimate large wave heights. Be that as it may, the maximum wave height scientifically recorded thus far (in 1965) in the North Atlantic was 67 feet by a wave recorder aboard an ocean station weather vessel, but this does not mean that higher waves do not occur.

Maximum Wave Height and Fetch

For a given wind velocity, the wave height becomes greater the longer the stretch of water (fetch) over which the wind has blown. Even with a very strong wind, the wave height for a given fetch does not exceed a certain maximum value.

Wave Speed and Fetch

At a given wind speed, the wave speed increases with increasing fetch.

Wave Height and Wind Speed

The height in feet of the greatest waves with high wind speeds has been observed to be about 0.8 of the wind speed in knots.

Wave Speed and Wind Speed

Although the ratio of wave speed to wind speed has been observed to vary from less than 0.1 to nearly 2.0, the average maximum wave speed apparently exceeds slightly the wind speed when the latter is less than about 25 knots, and is somewhat less than the wind speed at higher wind speeds.

Wave Height and Duration of Wind

The time required to develop waves of maximum height corresponding to a given wind increases with increasing wind speed. Observations show that with strong winds, high waves will develop in less than 12 hours.

Wave Speed and Duration of Wind

Although observational data are inadequate, it is known that for a given fetch and wind speed, the wave speed increases rapidly with time.

Wave Steepness

No well-established relationship exists between wind speed and wave steepness. This lack is probably due to the fact that wave steepness is not directly related to the wind speed, but depends upon the stage of development of the wave. The stage of development, or *age of the wave*, can be conveniently expressed by the ratio of the wave speed to the wind speed (C/U), because during the early stages of their formation, the waves are short and travel with a speed much less than that of the wind, while at later stages the wave speed may exceed the wind speed.

Decrease of Height of Swell

The height of swell decreases as the swell advances. Roughly, the swells lose one-third of their height each time they travel a distance in miles equal to their wave length in feet.

Increase of Period of Swell

Some scientists claim that the period of the swell remains unaltered when the swell advances from the generating area, whereas others claim that the period increases. The greater amount of evidence at the present time indicates that the period of the swell increases as the swell advances.

Table 11–1   **Probable Maximum Heights of Waves with Various Wind Speeds and Unlimited Fetch**

Wind Speed (Knots)	Wave Height (feet)
8	3
12	5
16	8
19	12
27	20
31	25
35	30
39	36
43	39
47	45
51	51

Table 11–2   **Wave Heights (Feet) Produced by Different Wind Speeds Blowing for Various Lengths of Time**

Wind Speed (Knots)	Duration (Hours)						
	5	10	15	20	30	40	50
10	2	2	2	2	2	2	2
15	4	4	5	5	5	5	5
20	5	7	8	8	9	9	9
30	9	13	16	17	18	19	19
40	14	21	25	28	31	33	33
50	19	29	36	40	45	48	50
60	24	37	47	54	62	67	69

Table 11–3   Wave Heights (Feet) Produced by Different Wind Speeds Blowing Over Different Fetches

Wind Speed (Knots)	Fetch (Nautical Miles)					
	10	50	100	300	500	1,000
10	2	2	2	2	2	2
15	3	4	5	5	5	5
20	4	7	8	9	9	9
30	6	13	16	18	19	20
40	8	18	23	30	33	34
50	10	22	30	44	47	51

Table 11–4   Minimum, Average, and Maximum Wave Heights (Feet) for the Tropical Trade-Wind Belts

Area	Minimum	Average	Maximum
Atlantic	0	6	20
Pacific	0	10	25
Indian	3	9	16

Table 11–5   Maximum Wave Heights with Various Wind Speed and the Fetches and Durations Required to Produce Waves 75 percent as high as the Maximum with each Wind Speed

Wind Speed (Knots)	Maximum Wave Height (Feet)	75 Percent of Maximum Height (Feet)	Fetch For 75 Percent (N.M.)	Duration for 75 Percent (Hours)
10	2	1.5	13	5
20	9	6.8	36	8
30	19	14.3	70	11
40	34	25.5	140	16
50	51	38.3	200	18

Table 11–6   Average Wave Length Compared to Wind Speed

Average Wave Length (Feet)	Wind Speed (Knots)
52	11
124	20
261	30
383	42
827	56

Tables 11–1 through 11–6 have been included to enable the reader to prepare his own preliminary oceanographic forecasts in conjunction with available weather data and weather maps.

SIGNIFICANT WAVES

Because of the irregular appearance of the sea surface, it is necessary to describe the waves that are present by means of some statistical term. This term should give emphasis to the higher waves because they are operationally more important than the smaller ones, although the actual number of smaller (and shorter) waves may be greater. For this reason, it is not advisable to state the mean wave height during, for example, a half-hour or one-hour period of observation, but rather to use the average height of the *highest one-third* of all observed waves.

We shall use this average and call it the *significant* wave height. This measure, as well as the mean, is not an exact measure because it depends upon the extent to which small waves have been recorded. If every ripple is counted as a wave, both the mean height and the average height of the highest one-third waves are reduced. In practice, all waves less than one foot are eliminated from consideration. Tests have indicated that the average height, rather than the mean height, of the highest one-third of all observed waves is a more consistent measure. This occurs because a casual observer tends to pay more attention to the higher waves and reports a wave height that lies closer to the significant wave height than to the mean wave height. The average height also depends less on the scope of the observations than does the mean height.

Table 11–7 shows the wave height characteristics of this measure. The significant wave height is given a relative value of 1.00. Therefore, if the significant wave height is known, the height of the maximum wave, the average height of the highest 10 percent, and the average height of the entire wave train can be computed.

By using table 11–7, it is seen, for example, that if a wave train has a significant wave height of 10 feet, the highest wave is 18.7 feet, the average of the highest 10 percent is 12.9 feet, and the mean wave height is 6.4 feet.

Table 11–7   A Comparison of Wave-Height Characteristics

Wave Terminology	Relative Height
Significant	1.00
Average	0.64
Highest 10 Percent	1.29
Highest	1.87

ENERGY OF WAVES

A train of waves in the ocean has potential energy due to the elevation and depression of the surface from its initial state of being level, and also has kinetic energy due to the movement of every particle in a vertical circular orbit, which we discussed earlier. Theoretical reasoning shows that the amounts of potential and kinetic energy are equal and proportional to the wave length times the wave height squared (LH^2) per wave length per unit of crest length. Thus, the kinetic energy in waves is tremendous. A four-foot, 10-second wave striking a coast expends more than 35,000 horsepower per mile of coast!

The power of waves can best be visualized by viewing the damage they cause. On the coast of Scotland, a block of cemented stone weighing 1,350 tons was broken loose and moved by waves. Five years later, the replacement pier, weighing 2,600 tons, was carried away. Engineers have measured the force of breakers along this coast of Scotland at 6,000 pounds per square foot!

Off the coast of Oregon, the roof of a lighthouse 91 feet above low water was damaged by a rock weighing 135 pounds.

An attempt has been made to harness the energy of waves along the Algerian coast. Waves are funneled through a V-shaped concrete structure into a reservoir. The water flowing out of the reservoir operates a turbine which generates power.

The energy of deep-water waves moves in their direction of travel with only half of the wave speed. This fact is responsible for the propagation of wave trains with only half the velocity of the individual waves. In any group of waves, the ones in the middle have the largest amplitude, or height, while those at the front and rear have small amplitudes which decrease to almost nothing at the edges of the group. While the group is traveling, the waves are continually overtaking the front of the wave train where they disappear, and thus the group velocity of a particular group of waves is one-half the velocity of the individual waves.

Waves lose energy when they break at the crest with the formation of "white horses," or when the wind begins to blow with a component from the opposite direction. Some of their energy is consumed in overcoming the internal friction (known as *viscosity*) which is evident between masses of water moving relative to each other. Molasses, for example, is a very viscous substance in which it is difficult to produce wave motion. The viscosity of water, however, is small, and this means that wave trains can travel great distances before they are finally dissipated.

OPTIMUM TRACK SHIP ROUTE FORECASTING

The shortest route between two points on our globe (as every seaman knows) is a great-circle track, from the standpoint of distance alone. But because of wind and sea conditions, it is seldom the shortest in time or the safest or most comfortable. In one year (1954), more than six percent of the entire world's shipping experienced heavy weather damage, and more than three percent was involved in collisions caused by weather and/or sea conditions.

In the early 1950s, the U.S. Navy established a ship routing service, which is an efficient and modern version of the service provided by Lieutenant Matthew Fontaine Maury, USN, before the Civil War, when he gathered the logs of ships and produced charts of ocean currents and winds. Maury's work resulted in saving days or weeks in the journeys of sailing vessels. Now, time savings are measured in hours and days.

The basic principles of ship routing are simultaneously simple and complex. Marine meteorologists predict wind speeds and directions for the particular areas of interest. Charts are prepared to show lines of equal wind velocity (isotachs), and these are then translated into charts which depict forecast wave heights, as shown in figure 11–9. By use of these charts, maximum attainable and safe speed can be computed for any type of ship or vessel. A ship utilizing this service maintains communications with the organization supplying the ship routing service and receives a daily course to be steered. Many millions of dollars are now saved each year in faster transits, less fuel consumed, minimizing storm damage, and so forth.

Although optimum track ship route forecasting was developed by the U.S. Navy, civilian private forecasters and organizations now furnish the same service to commercial ships and private individuals.

DETERMINATION OF STORM DIRECTION AND DISTANCE BY OBSERVING SWELL

If the height and period of the swell reaching you are observed, it is possible to determine *approximate* values of the distance of the storm (or more accurately, the end of the wave-generating area) from which the swell came; the travel time of the swell (the length of time it took the swell to

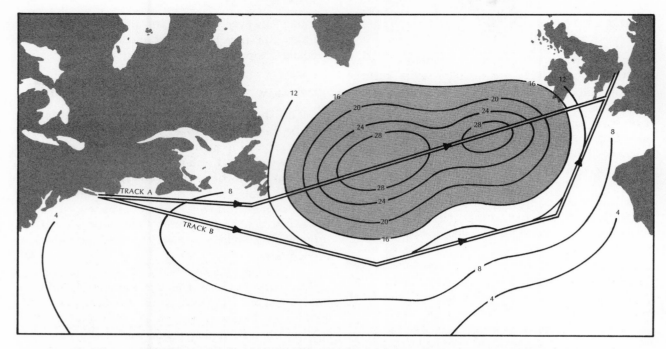

Figure 11–9. Optimum track ship routing across the Atlantic. Track A is the great circle route from New York to Southampton, England. Track B is the optimum ship track recommended by marine meteorologists in this weather/oceanographic situation. Contours are predicted wave heights in feet. Shading indicates areas of waves in excess of 16 feet. Although Track A is the shortest distance between New York and Southampton, ships following it would encounter 28-foot waves, would have to slow their speed of advance considerably, would take longer to arrive at their destination, and would probably sustain some storm damage either to the ship or cargo or both. On the recommended Track B, maximum wave heights would be only 12 feet, and the voyage would be made in less time and with no storm damage.

travel from the storm to your position); and the wind speed in the storm (or generating area) that created the swell. This is most important information to have, in order that a sound decision may be made regarding storm evasion, other courses of action, etc.

The values which can be derived from figure 11–10, it must be re-emphasized, are only approximate, because the height and period of the swell depend also upon the relation between wave and wind speeds (C/U) at the end of the fetch (the generating area of the waves). For our purpose, the ratio is assumed to vary in accordance with an assumed relationship between wind speed and duration (the length of time the wind has blown). In figure 11–10, two specific relationships between wind speed and duration are shown in the inset of the diagram. In the upper and lower parts of the diagram, corresponding values of decay distance, travel time, and wind speed are shown.

Choice of the better relationship between wind speed and duration must be based on a knowledge of weather situations which prevail in the area under consideration, and on the common experience

that high winds are usually of relatively short duration, while weaker winds may blow for a long period of time. For some combinations of observed swell height and period, only one of the two parts of figure 11–10 will apply.

As mentioned earlier, one must realize that the results obtained from figure 11–10 are approximate, since the wind speed and duration relationship may vary considerably from one weather situation to another. A lack of complete knowledge regarding the changes caused by following or opposing winds, and inaccuracies in the observations of the swell height and period will also introduce errors, though small. The values of distance and travel time obtained from figure 11–10 are more accurate than values of wind speed. We will try two examples to ensure our knowledge of how to use the diagram.

CASE 1: As we cruise along at greatly reduced speed, we are subjected to the effects of a 20-foot swell with a 12-second period, coming from a direction of 070 degrees true. What is the distance of the storm (or generating area) from our vessel? What is the bearing of the storm from our position? How long did it take the swells to reach our posi-

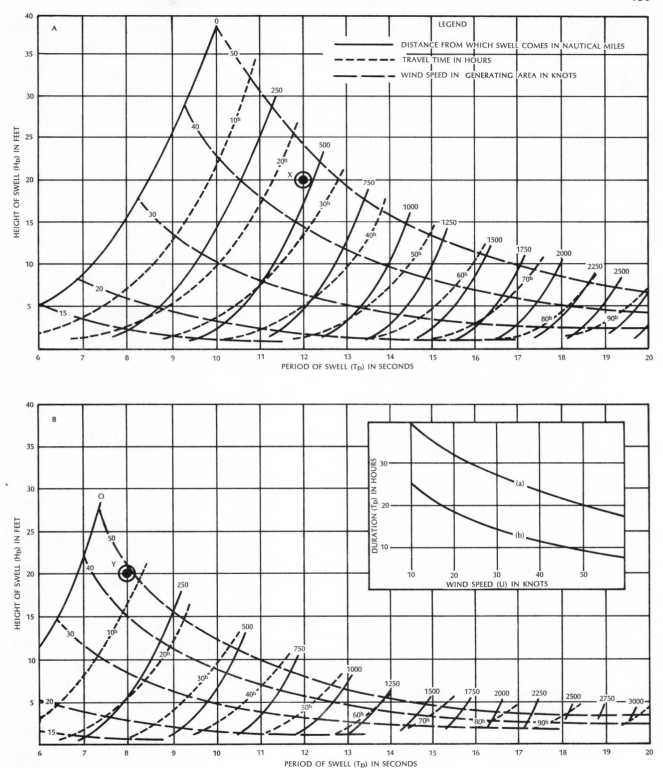

Figure 11–10. Diagrams to determine the distance from which a swell comes, the travel time of the swell, and the wind speed in the generating area, as functions of the observed height and period of the swell.

tion? What was the wind speed that generated the swells now causing us this discomfort?

Referring to figure 11–10, in the upper diagram, the dot marked X is plotted at the intersection of the horizontal line labelled *20 feet,* and the vertical line labelled *12 seconds.* The three sets of curves on the diagram provide us with three of the answers to this problem. Interpolating between the three sets of curves, (distance from which the swell comes, travel time, and wind speed in the generating area or storm), we obtain the following answers:

1. The storm (or generating area) is 440 nautical miles away from our position.
2. At the time the swell was generated, the storm (or generating area) was on a bearing of 070 degrees true from our position. (But remember, it will have moved in the time interval between generation and arrival at our position).
3. It took the swells 24 hours to reach our position.
4. Wind speeds in the storm (or generating area) which created the swells were 46 knots.

CASE 2: In the same sort of circumstances as mentioned in Case 1, we observe a 20-foot swell with an 8-second period, coming from a direction of 135 degrees true. Again, what is the distance of the storm (or generating area) from our vessel? What is the bearing of the storm from our position? How long did it take the swells to reach our position? What was the wind speed which generated the swells?

Referring to figure 11–10, again, in the lower diagram, the dot marked Y plotted at the intersection of the horizontal line labelled *20 feet,* and the vertical line labelled *8 seconds* represents the observed conditions of the swell. Again, interpolating between the three sets of curves in the lower diagram, we obtain the following answers:

1. The storm (or generating area) is 120 nautical miles distant from our position.

2. At the time the swell was generated, the storm (or generating area) was on a bearing of 135 degrees true from our position.
3. It took the swells eight hours to reach our position.
4. Wind speeds in the storm (or generating area) that created the swells were 47 knots.

OTHER BENEFITS FROM A KNOWLEDGE OF WAVES

At the present time, a great deal remains to be learned regarding the generation, travel, and decay of ocean waves and swell. However, we do know quite a bit about these phenomena at the present state of the art. A knowledge of the height and other characteristics of waves is of considerable practical value for a variety of purposes, including the design and behavior of vessels at sea; the design and orientation of harbors and the construction of breakwaters; problems of coast erosion and silting; the discharging of ships in open anchorages; and naval amphibious and carrier operations. For example, the behavior of an individual vessel or small craft at sea is governed to a considerable extent by the period of her roll and pitch in various conditions of loading in relation to the period of the waves she encounters, and her longitudinal and transverse strength calculations must inevitably take into account similar factors.

As mentioned in chapter 8, in tropical and subtropical waters, the arrival of a rather gentle swell (if from an unusual direction) may well be the first warning of the approach of a dangerous hurricane or typhoon. Similarly, wave data have value in areas where direct observation of storms are not available because of the limited amount of shipping in the area. From time to time, public investigations and inquiries are held into shipping losses where the destruction of life and property was serious, and in such cases, evidence provided by actual observations of wave or swell height from vessels in the vicinity has always proved to be of great value.

Chapter Twelve

Simple Forecasting Methods and Rules

"When the wind backs
And the weatherglass falls,
Then be on your guard
Against gales and squalls."
—Source unknown

In order that professional and amateur weathermen alike may understand, follow, and predict the weather, detailed observations must be taken *at the same time* over extensive regions, such as the entire northern hemisphere, and exchanged internationally. When an observer on the east coast of the United States makes a weather observation at 7:00 a.m. Eastern Standard Time, an observer on the Pacific Coast makes his at 4:00 a.m. Pacific Standard Time, and an observer in London, England makes his weather observation at noon Greenwich Time. These simultaneous observations provide the *synoptic* weather picture, the state of the earth's atmosphere at that particular instant of time. Surface observations are made around the world at least every six hours at 0000, 0600, 1200, and 1800 Greenwich Time, and upper-air observations are made at least every 12 hours at 0000 and 1200 Greenwich Time. More frequent observations are made when required or in special circumstances.

This wealth of information must be presented visually in such form that the users of weather charts can determine the current situation almost at a glance, perform the various analyses, and prepare appropriate forecasts. The pictorial presentation of weather data on the different charts serves this purpose.

HOW TO READ WEATHER MAPS

The information at each station of the weather networks of the world (and of ships at sea), whether surface or upper-air observations, is arranged around a circle drawn at the location of the station (and current position of the ships at sea). Analyses are performed depending upon the particular purpose of the chart.

The following weather information is plotted for each station and ship on a surface weather chart, and each item is always plotted in exactly the same position relative to the station/ship circle: wind direction and speed, pressure, temperature, dew point, visibility, ceiling, current weather (rain, snow, fog, etc.), the amount and types of clouds and their heights, pressure changes in the past three hours, weather in the past six hours, and the amount and type of precipitation. This tremendous amount of information for each station (and ship) is shown by symbols and numbers around the circle in a small space easily covered by a dime. Figures 12–1 and 12–2 are examples of land-station and ship weather plots, respectively, as they appear on surface weather maps. The relative position of each weather element around the circle should be memorized by the reader. Figure 12–3 contains all the weather symbols and notations used by weathermen in all countries of the world by international agreement. The reader is not expected to memorize all these symbols—though meteorologists must—but should keep the chart handy for reference when working with weather maps.

WIND SPEED
WIND DIRECTION
TYPE OF MIDDLE CLOUD
BAROMETRIC PRESSURE REDUCED TO SEA LEVEL
TEMPERATURE
PRESSURE HIGHER OR LOWER THAN 3 HOURS AGO
AMOUNT OF BAROMETRIC CHANGE IN LAST 3 HOURS
250
PRESENT WEATHER
BAROMETRIC TENDENCY IN LAST 3 HOURS
31
VISIBILITY
24
28
30
6
4
DEW POINT
2
45
TIME PRECIPITATION BEGAN OR ENDED
THE WEATHER DURING PAST 6 HOURS
AMOUNT OF PRECIPITATION DURING PAST 6 HOURS
TYPE OF LOW CLOUD
AMOUNT OF LOW CLOUDS
BASE HEIGHT OF LOW CLOUDS

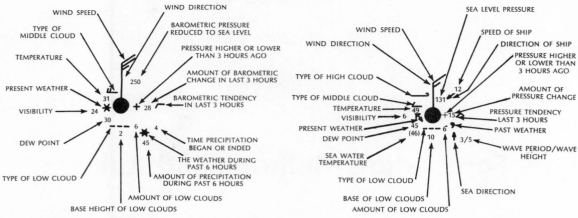

Figure 12–1. Land station weather plot

SEA LEVEL PRESSURE
WIND SPEED
SPEED OF SHIP
WIND DIRECTION
DIRECTION OF SHIP
PRESSURE HIGHER OR LOWER THAN 3 HOURS AGO
TYPE OF HIGH CLOUD
12
131
AMOUNT OF PRESSURE CHANGE
TYPE OF MIDDLE CLOUD
TEMPERATURE
49
PRESSURE TENDENCY LAST 3 HOURS
VISIBILITY
6
+15
PRESENT WEATHER
45
PAST WEATHER
DEW POINT
(46)
10
6
3/5
WAVE PERIOD/WAVE HEIGHT
SEA WATER TEMPERATURE
TYPE OF LOW CLOUD
SEA DIRECTION
BASE OF LOW CLOUDS
AMOUNT OF LOW CLOUDS

Figure 12–2. Ship weather plot

00 Cloud development NOT observed or NOT observable during past hour.§
01 Clouds generally dissolving or becoming less developed during past hour.§
02 State of sky on the whole unchanged during past hour.§
03 Clouds generally forming or developing during past hour.§
04 Visibility reduced by smoke.
05 Haze.
06 Widespread dust in suspension in the air, NOT raised by wind, at time of observation.
07 Dust or sand raised by wind, at time of ob.
08 Well develope devil(s) within p
10 Light fog.
11 Patches of shallow fog at station, NOT deeper than 6 feet on land.
12 More or less continuous shallow fog at station, NOT deeper than 6 feet on land.
13 Lightning visible, no thunder heard.
14 Precipitation within sight, but NOT reaching the ground.
15 Precipitation within sight, reaching the ground, but distant from station.
16 Precipitation within sight, reaching the ground, near to but NOT at station.
17 Thunder heard, but no precipitation at the station.
18 Squall(s) with during past hour
20 Drizzle (NOT freezing and NOT falling as showers) during past hour, but NOT at time of ob.
21 Rain (NOT freezing and NOT falling as showers) during past hr., but NOT at time of ob.
22 Snow (NOT falling as showers) during past hr., but NOT at time of ob.
23 Rain and snow (NOT falling as showers) during past hour, but NOT at time of observation.
24 Freezing drizzle or freezing rain (NOT falling as showers) during past hour, but NOT at time of observation.
25 Showers of rain during past hour, but NOT at time of observation.
26 Showers of snow, or of rain and snow, during past hour, but NOT at time of observation.
27 Showers of hail, or of hail and rain, during past hour, but NOT at time of observation.
28 Fog during pas but NOT at time
30 Slight or moderate duststorm or sandstorm, has decreased during past hour.
31 Slight or moderate dustorm or sandstorm, no appreciable change during past hour.
32 Slight or moderate dustorm or sandstorm, has increased during past hour.
33 Severe duststorm or sandstorm, has decreased during past hr.
34 Severe duststorm or sandstorm, no appreciable change during past hour.
35 Severe duststorm or sandstorm, has increased during past hour.
36 Slight or moderate drifting snow, generally low.
37 Heavy drifting snow, generally low.
38 Slight or m drifting snow, ge high.
40 Fog at distance at time of ob., but NOT at station during past hour.
41 Fog in patches.
42 Fog, sky discernible, has become thinner during past hour.
43 Fog, sky NOT discernible, has become thinner during past hour.
44 Fog, sky discernible, no appreciable change during past hour.
45 Fog, sky NOT discernible, no appreciable change during past hour.
46 Fog, sky discernible, has begun or become thicker during past hr.
47 Fog, sky NOT discernible, has begun or become thicker during past hour.
48 Fog, depositin sky discernible.
50 Intermittent drizzle (NOT freezing) slight at time of observation.
51 Continuous drizzle (NOT freezing) slight at time of observation.
52 Intermittent drizzle (NOT freezing) moderate at time of ob.
53 Continuous drizzle (NOT freezing), moderate at time of ob.
54 Intermittent drizzle (NOT freezing), thick at time of observation.
55 Continuous drizzle (NOT freezing), thick at time of observation.
56 Slight freezing drizzle.
57 Moderate or thick freezing drizzle.
58 Drizzle and slight
60 Intermittent rain (NOT freezing), slight at time of observation.
61 Continuous rain (NOT freezing), slight at time of observation.
62 Intermittent rain (NOT freezing), moderate at time of ob.
63 Continuous rain (NOT freezing), moderate at time of observation.
64 Intermittent rain (NOT freezing), heavy at time of observation.
65 Continuous rain (NOT freezing), heavy at time of observation.
66 Slight freezing rain.
67 Moderate or heavy freezing rain.
68 Rain or drizz snow, slight.
70 Intermittent fall of snow flakes, slight at time of observation.
71 Continuous fall of snowflakes, slight at time of observation.
72 Intermittent fall of snow flakes, moderate at time of observation.
73 Continuous fall of snowflakes, moderate at time of observation.
74 Intermittent fall of snow flakes, heavy at time of observation.
75 Continuous fall of snowflakes, heavy at time of observation.
76 Ice needles (with or without fog).
77 Granular snow (with or without fog).
78 Isolated starlik crystals (with or w fog).
80 Slight rain shower(s).
81 Moderate or heavy rain shower(s).
82 Violent rain shower(s).
83 Slight shower(s) of rain and snow mixed.
84 Moderate or heavy shower(s) of rain and snow mixed.
85 Slight snow shower(s).
86 Moderate or heavy snow shower(s).
87 Slight shower(s) of soft or small hail with or without rain or rain and snow mixed.
88 Moderate or shower(s) of soft o hail with or witho or rain and snow
90 Moderate or heavy shower(s) of hail††, with or without rain or rain and snow mixed, not associated with thunder.
91 Slight rain at time of ob.; thunderstorm during past hour, but NOT at time of observation.
92 Moderate or heavy rain at time of ob.; thunderstorm during past hour, but NOT at time of observation.
93 Slight snow or rain and snow mixed or hail† at time of observa.; thunderstorm during past hour, but not at time of observation.
94 Mod. or heavy snow, or rain and snow mixed or hail† at time of ob.; thunderstorm during past hour, but NOT at time of observation.
95 Slight or mod. thunderstorm without hail†, but with rain and/or snow at time of ob.
96 Slight or mod. thunderstorm, with hail† at time of observation.
97 Heavy thunderstorm, without hail†, but with rain and/or snow at time of observation,
98 Thunderstorm bined with dusto sandstorm at time

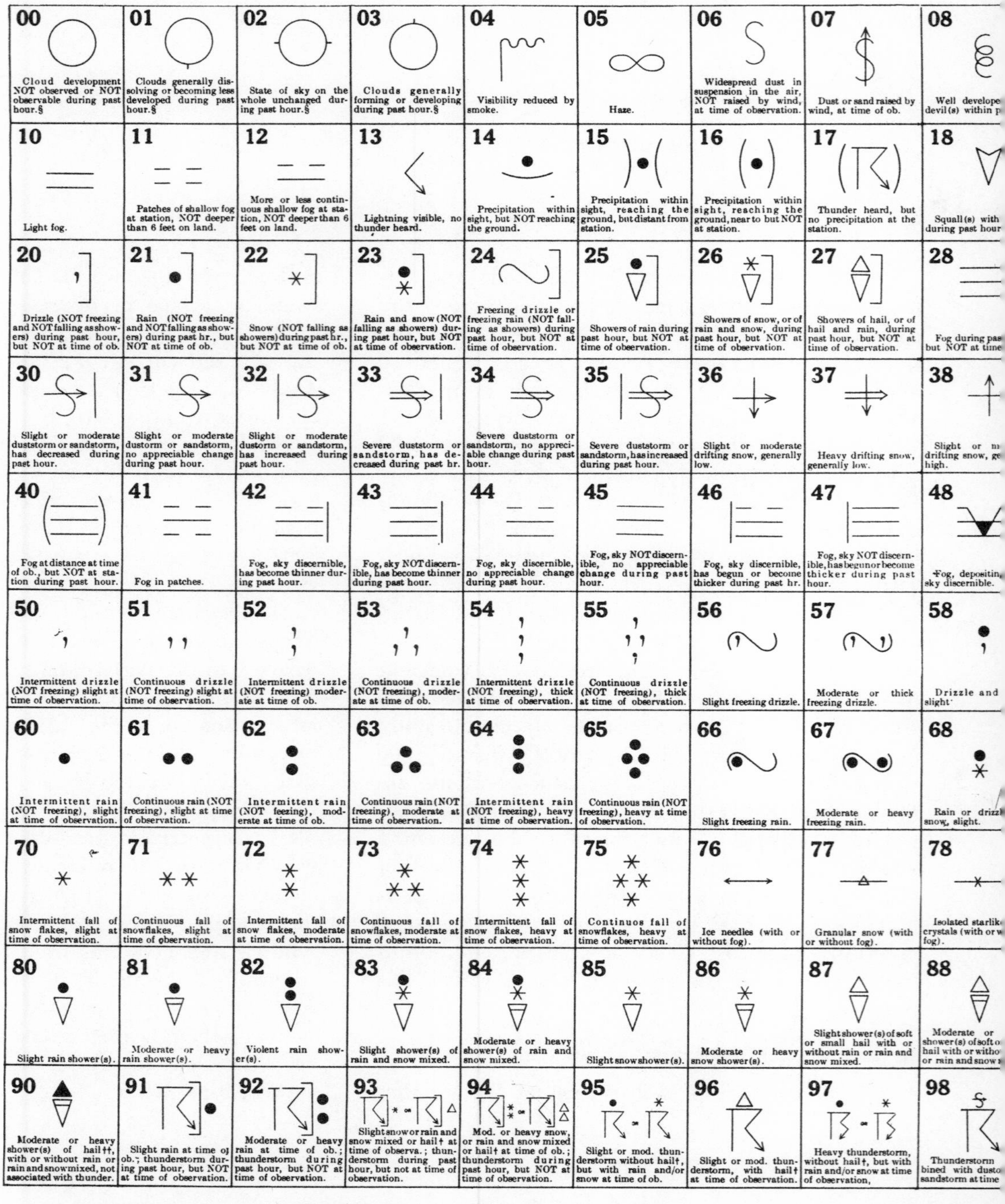

Figure 12–3. Weather code figures and symbols with their meanings

SYMBOLIC FORM OF SYNOPTIC CODE: iii Nddff VVwwW PPPTT $N_hC_LhC_MC_H$ T_dT_dapp $7RRR_ts$ $8N_sCh$

Daily weather maps of various types are available from a variety of sources—the ESSA Weather Bureau, most daily newspapers, private weather consultants, industrial meteorological organizations, etc. Figure 12–4 includes examples of ESSA Weather Bureau daily weather maps which are available at a nominal cost from the Superintendent of Documents, Government Printing Office, Washington, D.C., 20402. The surface weather map presents station data and the analysis for 0700 EST. The tracks of well-defined low pressure areas are indicated by chains of arrows (when applicable). The locations of these centers at the times 6, 12, and 18 hours preceding map time are indicated by small black squares enclosing white crosses. Areas of precipitation are indicated by shading. The weather reports printed on these maps are only a small fraction of those that are included in the operational weather maps, and on which analyses are based. Occasional apparent discrepancies between the printed station data and the analyses result from those station reports that cannot be included in the published maps because of lack of space. The symbols of fronts, pressure systems, air masses, etc., are the same as discussed in previous chapters. Note how the isobars are V-shaped (they "kink") at the

#	Present weather (ww)	Low clouds (CL)	Middle clouds (CM)	High clouds (CH)	Cloud type	Period / weather	Cloud amount	Pressure tendency (a)
0	m or sandstorm ...ght of or at sta-...ng past hour.	No Sc, St, Cu, or Cb clouds.	No Ac, As or Ns clouds.	No Ci, Cc, or Cs clouds.	Ci	Cloud covering ½ or less of sky throughout the period.	No clouds.	Rising then falling. Now higher than, or the same as, 3 hours ago.
1	cloud(s) with-...during past	Ragged Cu, other than bad weather, or Cu with little vertical develop-ment and seemingly flattened, or both.	As, the greatest part of which is semitrans-parent through which the sun or moon may be faintly visible as through ground glass.	Filaments, strands, or hooks of Ci, not in-creasing.	Cc	Cloud covering more than ½ of sky during part of period and cover-ing ½ or less during part of period.	One-tenth or less, but not zero.	Rising, then steady; or rising, then rising more slowly. Now higher than 3 hours ago.
2	erstorm (with ...t precipitation) ...past hour, but ...time of ob.	Cu of considerable de-velopment, generally towering, with or with-out other Cu or Sc; bases all at same level.	As, the greatest part of which is sufficiently dense to hide the sun or moon, or Ns.	Dense Ci in patches or twisted sheaves, usually not increasing; or Ci with towers or battle-ments or resembling cumuliform tufts.	Cs	Cloud covering more than ½ of sky through-out the period.	Two- or three-tenths.	Rising (steady or un-steadily). Now higher than 3 hours ago.
3	drifting snow, ...high.	Cb with tops lacking clear-cut outlines, but are clearly not fibrous, cirriform, or anvil-shaped; Cu, Sc, or St may be present.	Ac (most of layer is semitransparent) other than crenelated or in cumuliform tufts; cloud elements change but slowly with all bases at a single level.	Ci, often anvil-shaped, derived from or asso-ciated with Cb.	Ac	Sandstorm, or dust-storm, or drifting or blowing snow.	Four-tenths.	Falling or steady, then rising; or rising then rising more rapidly. Now higher than 3 hours ago.
4	epositing rime, ...discernible.	Sc formed by spread-ing out of Cu; Cu may be present also.	Patches of semitrans-parent Ac which are at one or more levels; cloud elements are continu-ously changing.	Ci, hook-shaped and/or filaments, spreading over the sky and gener-ally becoming denser as a whole.	As	Fog, or thick haze.	Five-tenths.	Steady. Same as 3 hours ago.
5	and rain, mod-...heavy.	Sc not formed by spreading out of Cu.	Semitransparent Ac in bands or Ac in one more or less continuous layer gradually spread-ing over sky and usually thickening as a whole; the layer may be opaque or a double sheet.	Ci, often in converg-ing bands, and Cs or Cs alone but increasing and growing denser as a whole; the continuous veil not exceeding 45° above horizon.	Ns	Drizzle.	Six-tenths.	Falling, then rising. Now lower than, or the same as, 3 hours ago.
6	or drizzle and ...d'te or heavy.	St in a more or less continuous layer and/or ragged shreds, but no Fs of bad weather.	Ac formed by the spreading out of Cu.	Ci, often in converg-ing bands, and Cs or Cs alone but increasing and growing denser as a whole; the continuous veil exceeds 45° above horizon but sky not totally covered.	Sc	Rain.	Seven-or eight-tenths.	Falling, then steady; or falling, then falling more slowly. Now lower than 3 hours ago.
7	ellets (sleet, ...finition).	Fs and/or Fc of bad weather (scud) usually under As and Ns.	Double-layered Ac or an opaque layer of Ac, not increasing over the sky; or Ac coexisting with As or Ns or with both.	Veil of Cs completely covering the sky.	St	Snow, or rain and snow mixed, or ice pel-lets (sleet).	Nine-tenths or more, but not ten-tenths.	Falling (steadily or unsteadily). Now lower than 3 hours ago.
8	shower(s) of ...with or with-... or rain and ...xed, not asso-...th thunder.	Cu and Sc (not formed by spreading out of Cu); base of Cu at a different level than base of Sc.	Ac with sprouts in the form of small towers or battlements or Ac hav-ing the appearance of cumuliform tufts.	Cs not increasing and not completely covering the sky.	Cu	Shower(s).	Ten-tenths.	Steady or rising, then falling; or falling, then falling more rapidly. Now lower than 3 hours ago.
9	thunderstorm ...t at time of ob.	Cb having a clearly fibrous (cirriform) top, often anvil-shaped, with or without Cu, Sc, St, or scud.	Ac, generally at sev-eral layers in a chaotic sky; dense Cirrus is usually present.	Cc alone or Cc accom-panied by Ci and/or Cs, but Cc is the predomi-nant cirriform cloud.	Cb	Thunderstorm, with or without precipitation.	Sky obscured, or cloud amount cannot be estimated.	Indicator figure. Re-gionally agreed elements and NOT "pp" are re-ported by the next two code figures.

$S_ps_ps_p$ $1d_wd_wP_wH_w$ $2h_{85}h_{85}h_{85}a_3$ $3R_{24}R_{24}R_{24}R_{24}$ $4T_xT_xT_nT_n$ (Additional Plain Language Data)

Figure 12–4.   ESSA Weather Bureau daily weather maps

fronts, as previously discussed.

The 500-*Millibar Chart* presents the height contours and isotherms of the 500-millibar surface at 0700 EST. The 500-millibar heights usually range from 15,800 to 19,400 feet. The height contours are shown as continuous lines and are labelled in feet above sea level. The isotherms are shown as dashed lines and are labelled in degrees Celsius. The arrows show the wind direction and speed at the 500-millibar level.

The *Highest and Lowest Temperature Chart* presents the maximum and minimum values for the 24-hour period ending at 0100 EST. The names of the reporting points can be obtained from the surface weather map. The maximum temperature is plotted above the station location, and the minimum temperature is plotted below this point.

The *Precipitation Areas and Amounts Chart* indicates by means of shading the areas that had precipitation during the 24 hours ending at 0100 EST. Amounts in inches to the nearest hundredth of an inch are for the same period. Incomplete totals are underlined. A *T* indicates only a trace of precipitation. Dashed lines show the depth of snow on the ground in inches as of 0700 EST of the previous day.

Figure 12–5 is an example of an American forecast (prognostic) surface weather map that appears

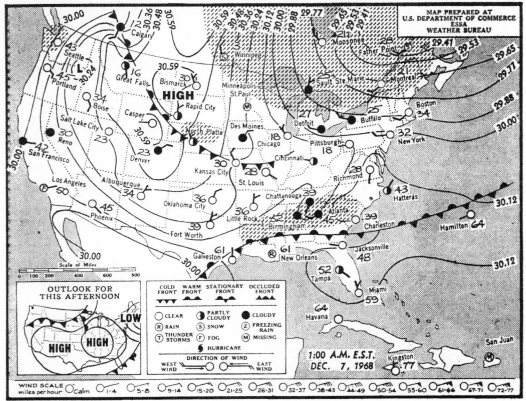

Figure beside Station Circle indicates current temperature (Fahrenheit); a decimal number beneath temperature indicates precipitation in inches during the six hours prior to time shown on map.

Cold front: a boundary line between cold air and a mass of warmer air, under which the colder air pushes like a wedge, usually advancing southward and eastward.

Warm front: a boundary between warm air and a retreating wedge of colder air over which the warm air is forced as it advances, usually northward and eastward.

Occluded front: a line along which warm air has been lifted by the action of the opposing wedges of cold air. This lifting of the warm air often causes precipitation along the front.

Shading on the map indicates areas of precipitation during six hours before time shown.

Isobars (solid black lines) are lines of equal barometric pressure and form pressure patterns that control air flow. Labels are in inches.

Winds are counter-clockwise toward the center of low-pressure systems, and clockwise and outward from high-pressure areas.

Pressure systems usually move eastward, averaging 500 miles a day in summer, 700 in winter.

Courtesy of the *New York Times*

Figure 12–5. Daily weather map in a United States newspaper

in *The New York Times,* and figure 12–6 illustrates its British counterpart that appears in *The Guardian.*

Although newspaper daily weather maps are not as detailed as the official charts of the ESSA U.S. Weather Bureau or foreign governmental meteorological organizations, they are timely, readily accessible, inexpensive, and can be used to great advantage by the professional mariner and weekend sailor with some knowledge of the weather science. At sea, weather maps obtained by radio facsimile serve the same purpose.

THE WEATHER: Bright spells and showers

A complex area of low pressure over the British Isles will move away to NW. Scotland and Northern Ireland will have a good deal of cloud, but there will be some bright intervals. Most parts will have rain or snow, and there are likely to be some longer outbreaks of rain or snow in N Scotland. N England will have sunny intervals and wintry showers, but some places may be rather misty at first. Wales and the rest of England will have bright spells and showers. They will be most frequent in W districts, and will be of sleet and snow at times, especially over high ground. More general rain or snow may spread into SW later in the day. Temperatures will be generally below normal with frost in many places early and late.

London area, SE and Cent S England, East Anglia, Midlands: Sunny intervals and scattered showers. Sleet at times. Wind W. moderate. Max. temp. 5C (41F).

E England, Cent N and NE England: Misty at first. Sunny intervals and scattered wintry showers later. Wind W. moderate. Max. 4C (39F).

Channel Islands, SW England, S Wales: Sunny intervals and wintry showers. Perhaps more general rain or snow later. Wind W, becoming SW, moderate. Max. 6C (43F).

N Wales, NW England, Lake District, Isle of Man, SW Scotland, Glasgow area, Argyll, N Ireland: Bright intervals and wintry showers. Wind W, moderate or fresh. Max. 5C (41F).

Borders, Edinburgh, E Scotland, Aberdeen area, Cent Highlands: Bright intervals and scattered wintry showers. Perhaps misty at first. Wind W. moderate. Max 5C (41F).

Moray Firth area, Caithness, NW Scotland, Orkney, Shetland: Rather cloudy. Occasional rain, sleet, or snow. Perhaps some bright intervals. Wind S to SW, moderate. Max 4C (39F).

Outlook: Continuing rather cold with rain at times, and some snow over high ground. Frost in many places at night.

SEA PASSAGES

S North Sea, Strait of Dover, English Channel (E): Moderate. St George's Channel, Irish Sea: Moderate to rough.

THE SATELLITES

The figures give in order: Time of visibility; where rising; maximum elevation and direction of setting. An asterisk indicates entering or leaving eclipse.

Pageos A, 18.17–18.41 N 85 SSE S and 21.19–21.33 NNW 25 WNW NNW.

LIGHTING-UP TIMES

Birmingham .... 5 55 p.m. to 8 40 a.m.
Bristol .... 6 03 p.m. to 8 37 a.m.
London .... 5 59 p.m. to 8 27 a.m.
Nottingham .... 5 51 p.m. to 8 39 a.m.

HIGH-TIDE TABLE

London Bridge 1 29 a.m. ... 1 59 p.m.
Dover ...... 11 04 a.m. ... 11 35 p.m.

SUN RISES ..... 8 59 a.m.
SUN SETS ...... 5 23 p.m.
MOON RISES .... 9 06 a.m.
MOON SETS ..... 3 59 p.m.
MOON: New Moon tomorrow.

LONDON READINGS

7 p.m. Wednesday to 7 a.m. yesterday: Temp. min. 2C (36F); rainfall .16in. 7 a.m. to 7 p.m. yesterday; Temp. max. 7C (45F); rainfall, nil: sunshine 1.1 hours.

Around Britain

Reports for the 24 hours ended 6 p.m. yesterday:

	Sunshine hrs.	Rain in.	Max. temp. C	Max. temp. F	Weather (day)
EAST COAST					
Scarboro	—	.23	4	40	Rain p.m.
Bridlingtn	—	.78	4	39	Rain
Lowestoft	—	.47	5	41	Rain
Clacton	1.5	.64	4	40	Cloudy
Whitstable	1.3	.64	6	42	Cloudy
Herne Bay	1.6	.51	5	41	Cloudy
SOUTH COAST					
Folkestone	1.8	.17	5	41	Cloudy
Hastings	2.0	.25	6	42	Cloudy
Eastbourne	3.9	.16	6	43	Sunny intls
Brighton	3.4	.39	6	43	Sunny intls
Worthing	3.9	.31	6	43	Sunny intls
Littlehptn	2.7	.40	7	44	Sunny intls
Bognor Reg.	4.4	.20	6	43	Sunny prds
Southsea	2.4	.50	6	43	Sunny intls
Sandown	2.2	.5.	6	43	Sunny intls
Shanklin	2.7	.66	6	43	Sunny intls
Ventnor	3.0	.50	6	42	Shwr
Bournemth	1.3	.11	6	42	Shwrs
Poole	2.0	.21	6	43	Shwrs
Swanage	3.5	.09	6	43	Sunny intls
Weymouth	—	—	8	47	Drizzle
Exmouth	2.3	.33	6	43	Sleet shwrs
Teignmth	2.2	.34	6	43	Shwrs
Torquay	5.0	.57	5	41	Sunny prds
Falmouth	6.3	.21	7	44	Sunny pds
Penzance	5.8	.21	8	46	Gales, sunny prds
Jersey	5.3	.50	8	47	Sunny prds
WEST COAST					
Douglas	—	.37	5	41	Rain, gales
Morecambe	—	—	2	36	Fog
Blackpool	—	.04	2	36	Sleet
Southport	—	.05	3	38	Fog, snow
Colwyn Bay	—	.17	6	42	Rain
Llandudno	—	.18	6	42	Rain
Anglesey	0.1	.47	6	43	Rain
Weston-s-M.	—	.16	4	40	Cloudy
Ilfracombe	—	.17	7	45	Rain, gales
Newquay	3.3	.13	6	43	Snny ins, gls
Scilly Isles	5.9	.05	8	46	Sunny prds
INLAND					
Ross-on-Wy	1.7	.07	5	41	Cloudy
SCOTLAND					
Lerwick	0.7	.28	6	43	Hail
Wick	—	.28	4	39	Rain
Stornoway	—	.15	6	43	Rain
Kinloss	0.1	.21	5	41	Rain
Dyce	—	.45	4	39	Rain
Tiree	0.1	.55	6	43	Rain
Leuchars	0.3	.20	5	41	Rain, snow
Glasgow	—	.06	5	41	Rain
Eskdalemuir	—	.01	3	37	Snow, rain
NORTHERN IRELAND					
Belfast	2.1	.05	5	41	Showers

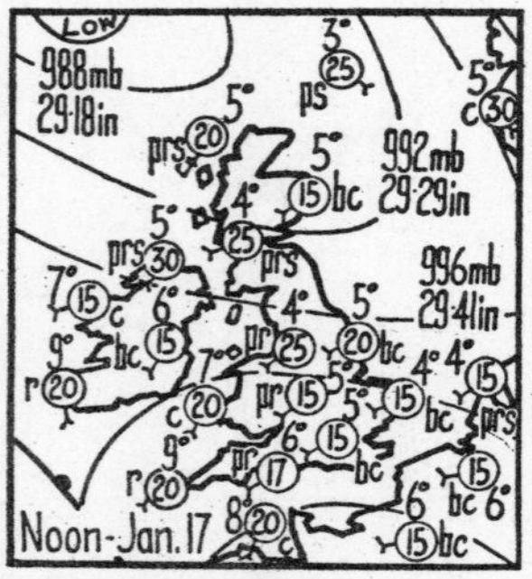

Around the world

Lunch-time reports

		C	F				C	F
Algiers	C	16	61		Lisbon	F	15	59
Amsterdm	C	6	43		Locarno	Sl	1	34
Athens	S	16	61		London	R	6	43
Barcelona	S	12	54		Luxembrg	C	4	39
Beirut	C	18	64		Madrid	F	10	50
Belfast	C	4	39		Majorca	S	14	57
Belgrade	S	11	52		Malaga	C	18	64
Berlin	C	2	36		Malta	C	15	59
Biarritz	C	9	48		Manchstr	Fo	0	32
Birminghm	R	3	37		Miami	Dr	3	37
Bristol	F	4	39		Moscow	C	-6	21
Brussels	C	5	41		Munich	S	-2	28
Budapest	C	3	37		Naples	R	11	52
Cardiff	C	3	37		Nice	S	11	52
Cologne	C	7	45		Nicosia	C	13	55
Copenhgn	Fo	2	36		Oslo	C	1	34
Dublin	C	1	34		Paris	R	4	39
Edinburgh	C	5	41		Prague	C	-1	30
Faro	C	16	61		Reykjavik	F	-6	21
Florence	R	8	46		Rome	Th	9	48
Frankfurt	C	2	36		Ronaldswy	R	5	41
Geneva	C	3	37		Stockholm	Dr	2	36
Gibraltar	F	17	63		Tel-Aviv	C	17	63
Guernsey	F	6	43		Toronto	C	-3	27
Helsinki	C	-1	30		Tunis	C	16	61
Innsbruck	C	0	32		Venice	C	8	46
Istanbul	C	12	54		Vienna	C	1	34
Jersey	F	-7	45		Warsaw	C	3	37
Las Palmas	C	19	66		Zurich	F	3	37

C, cloudy; Dr, drizzle; S, sunny; R, rain; Fo, fog; Sl, sleet; Th, thunderstorm.

Snow reports

	L	U	Piste	Off piste	Wthr	F
Andermatt	52	86	Good	Powder	Cloud	28
Davos	24	38	Good	Powder	Snow	36
Grindewald	14	39	Good	Fair	Cloud	31
Gstaad	10	30	Good	Powder	Snow	32
Lenzerheide	41	60	Good	Varied	Cloud	32
Murren	37	44	Good	Powder	Snow	30
Pontresina	45	80	Good	Powder	Fair	31
Sauze d'Oulx	18	60	Good	Powder	Fine	25
Verbier	18	36	Good	Varied	Snow	40
Zermatt	42	66	Good	Powder	Cloud	32

In the above reports, supplied by representatives of the Ski Club of Great Britain. L refers to lower slopes and U to upper slopes. The following reports have ben received from other sources.

	Depth in			
ITALY	L	U	Piste	Wthr	F
Bardonecchia	24	92	Good	Fine	23
Bormio	24	48	Good	Snow	29
Canazei	16	80	Good	Fine	23
Cervinia	48	100	Good	Fine	24
Claviere	36	60	Good	Fine	19
Cortina	73	120	Good	Fine	—
Courmayeur	22	92	Good	Fine	23
Macugnaca	20	120	Good	Cloud	29
Madesimo	60	160	Good	Cloud	23
Madonna di Campiglio	28	60	Good	Cloud	29
San Martino	12	44	Good	Fine	29
Selva	28	52	Good	Fine	29
GERMANY					
Berchtesgaden	10	36	Poor	Cloud	26
Garmisch	10	22	Poor	Fine	21
Hindelang	16	40	Poor	Cloud	21
Kleinwalsertal	22	40	Good	Fine	23
Mittenwald	8	52	Poor	Fine	26
Oberammergau	10	28	Poor	Fine	26
Oberjoch	16	40	Poor	Cloud	21
Oberstaufen	22	32	Poor	Fine	24
Oberstdorf	16	32	Poor	Fine	15
AUSTRIA					
Badgastein	10	35	Fair	Cloud	30
Berwang	9	30	Good	Cloud	30
Gallur	16	28	Good	Cloud	30
Mayrhofen	10	16	Good	Cloud	25
Obergurgl	30	55	Good	Cloud	25
Saalbach	8	12	Fair	Cloud	35
Schruns	21	32	Fair	Cloud	26
Solden	25	46	Good	Cloud	23
NORWAY					
		U	Piste	Wthr	
Finse		22	Good	Cloud	
Gol		28	Good	Snow	
Gello		28	Good	Cloud	
Voss		32	Good	Fair	
Lillehammer		38	Good	Snow	
Oslo		39	Good	Cloud	

SCOTLAND

Cairngorms: Main runs: General snow cover. New snow on a firm base. Lower slopes: General snow cover. New snow on a firm base. Maximum runs: 1,600ft. Access roads, slight snow.

Glenshee: Main runs: All runs complete. Powdery snow on a firm base. Lower slopes: Ample nursery areas. Powder snow on a firm base. Maximum runs: 1,000ft. Access roads: Clear.

Glencoe: Main runs: Some runs complete, others broken. Hard packed snow. Lower slopes: Ample nursery areas. Hard packed snow. Maximum: 1,100ft. Access roads: Clear.

Forecast: Snow showers, but also bright periods. Moderate W winds. Freezing level 500ft.

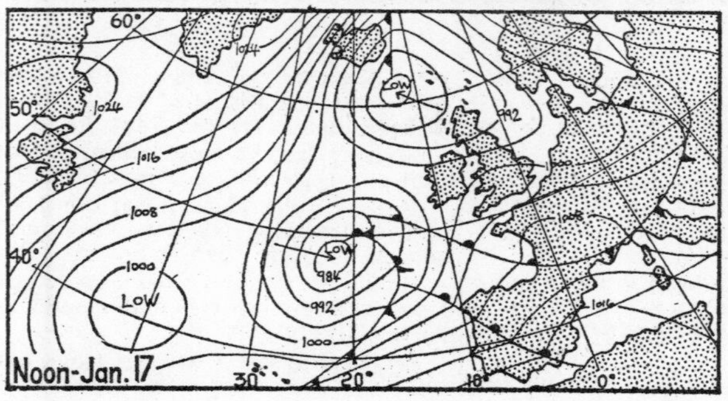

988mb 29.18in
992mb 29.29in
996mb 29.41in
Noon–Jan. 17

Noon–Jan. 17

Courtesy of Guardian Newspapers, Ltd.

Figure 12–6. Daily weather map and other information in a British newspaper

BEING YOUR OWN WEATHER FORECASTER
Persistence Method

Forecasting from a single weather map is not very difficult, but reasonable accuracy is limited to about 6 to 12 hours, at best. For this purpose, the best newspaper weather maps or the more complete ESSA Weather Bureau maps should be used. Short-range forecasting is based largely on the principle of *persistence*, which assumes, for example, that a front moving at 25 knots will continue at that speed with the same weather accompanying it; that a stationary high-pressure cell will remain stationary; that a fast-moving low-pressure system heading northeastward will continue moving in that direction at a rapid pace, and so forth. This principle is usually followed by even the expert forecasters for short-range periods. It generally works very well, though peculiar local conditions sometimes throw such predictions off.

Study the simplified weather map of figure 12–7. Note the continuous precipitation ahead of the warm front and the rainshowers behind the cold front. You are to prepare a forecast for Boston, Mass. At 1000 EST, the cold front located 150 nautical miles to the west of Boston is moving eastward at 25 knots. What is your forecast?

At the reported speed of 25 knots, the cold front will travel the 150-nautical-mile distance to Boston in six hours, and will arrive there at 1600 EST. Based on what you have learned thus far, your forecast for Boston should be: "Cold front passage accompanied by rainshowers at 1600 EST. Surface wind shifting from SW to NW during frontal passage. Cooler tonight and tomorrow."

Limitations of the persistence method of forecasting lie in the fact that so many things can change a weather pattern. For example, as the cold front of figure 12–7 sweeps across the eastern seaboard into the Atlantic, a brand-new secondary low-pressure system could develop to the south of Cape Hatteras, North Carolina (a favorite breeding area of secondary cyclones) and travel rapidly north-

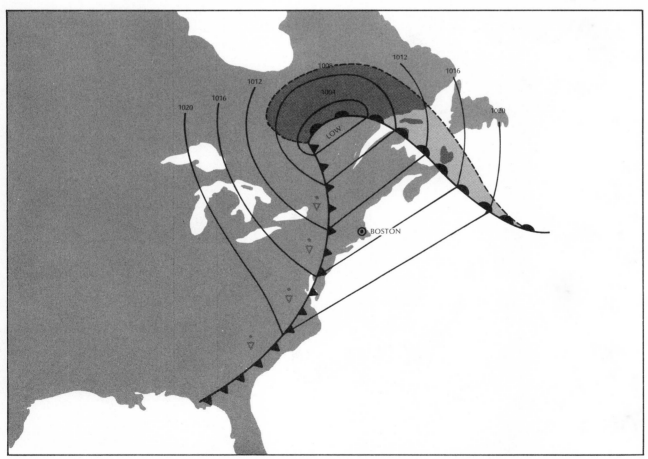

Figure 12–7. Making your own short-range weather forecast. It is now 1000 EST. At this time, the cold front is 150 nautical miles to the west of Boston and is moving eastward at 25 knots. What is your forecast for Boston?

eastward, paralleling the coastline. In this case, a typical "Nor'easter" would soon envelop the Boston area. A trained and well-experienced forecaster can accurately estimate and predict these developments, but on occasion, they are caught, too. A beginner in the weather business would not be likely to spot such a development in advance. So, keep your weather records and gain experience.

Continuity Method

The continuity method of simple forecasting is really the persistence method, enlarged and modified. It can be used with fair to good accuracy for periods much longer than 12 hours. One makes continuity forecasts by the intelligent use of a series of weather maps. Essentially, the method involves the extrapolation into the future of established trends—whether they be speed changes, intensity changes, directional movement changes, etc. Figure 12–8 illustrates, in a general way, the continuity method of forecasting. But let's try our hand at the continuity method, using an extremely simplified series of weather maps.

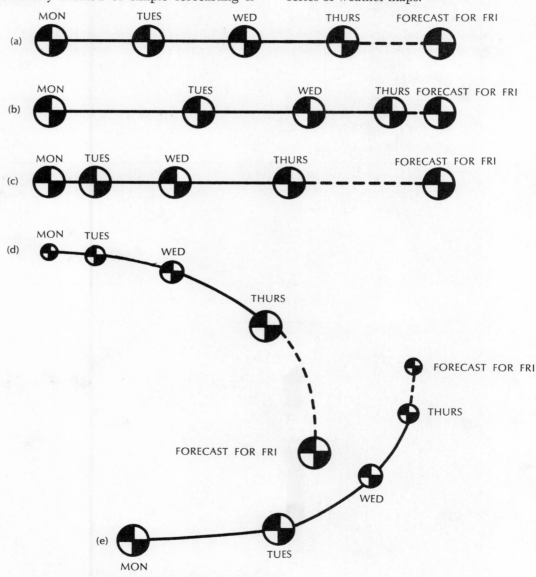

Figure 12–8.   The continuity method of forecasting.
    (a) Direction of movement in a straight line; speed of movement constant.
    (b) Direction of movement in a straight line; speed of movement decreasing.
    (c) Direction of movement in a straight line; speed of movement increasing.
    (d) Direction of movement curving to the right; speed of movement increasing.
    (e) Direction of movement curving to the left; speed of movement decreasing.

Figure 12–9 shows a 1008-millibar low-pressure system centered over the state of Colorado on a Monday, with a cold front extending to the southwest out of the low, and a warm front extending to the southeast. Rainshowers are occurring behind the cold front, and a steady rain is falling ahead of the warm front.

In figure 12–10, on Tuesday, the low has deepened by four millibars, with the central pressure now having a value of 1004 millibars. The low, itself, is now centered over extreme western Illinois, having traveled eastward at 30 knots, covering a distance of 720 nautical miles in the past 24 hours. The cold front is beginning to overtake the warm front, as evidenced by the warm sector (the area between the cold front and the warm front, south of the low center) becoming more narrow. The precipitation area ahead of the warm front becomes wider.

Figure 12–9. Simplified surface weather map for Monday

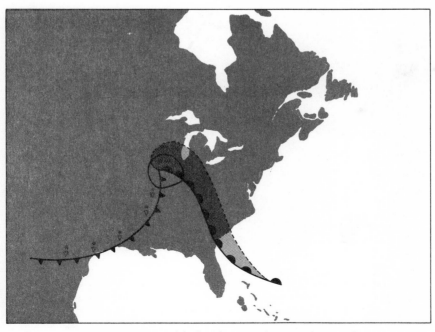

Figure 12–10. Simplified surface weather map for Tuesday

In figure 12–11, on Wednesday, the low has deepened by another four millibars, and the central pressure is now 1,000 millibars. The low is now centered over the extreme southwestern portion of the state of New York, having traveled east northeastward at 25 knots, covering a distance of 600 nautical miles in the past 24 hours. The cold front is now rapidly overtaking the warm front, and the frontal wave is just beginning to occlude. Steady precipitation is spreading farther in advance of the warm front.

In figure 12–12, on Thursday, the low-pressure system has deepened by an additional eight millibars, and the central pressure has fallen to 992 millibars. The low is now centered in extreme northern Maine, having curved slightly from an east northeast track to a northeast track and having decelerated further to an average speed of 20 knots in the

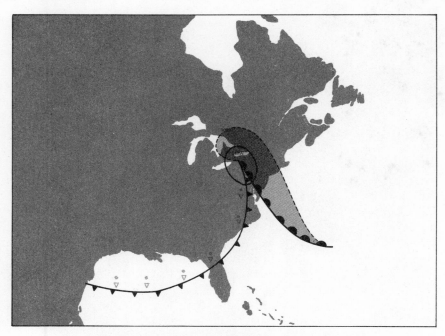

Figure 12–11. Simplified surface weather map for Wednesday

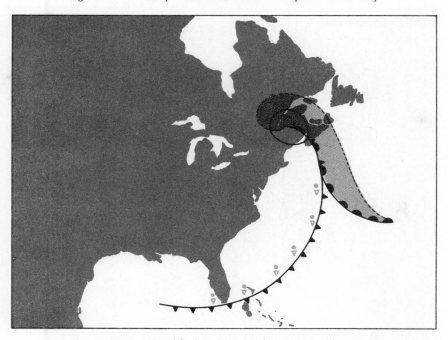

Figure 12–12. Simplified surface weather map for Thursday

past 24 hours, covering a distance of 480 nautical miles in the last 24-hour period. The cold air mass has now overtaken the warm air mass and formed an occluded front that extends from the center of the low to a position of latitude 40 degrees north, longitude 65 degrees west. The precipitation is much the same as before, except that the steady rain now extends over the top of the occlusion.

What would you expect Friday's weather map to look like? How would you draw your forecast (prognostic) chart for tomorrow (Friday)?

One of the simplest and best methods to do this is to take a blank map and place it over the Monday, Tuesday, Wednesday and Thursday maps in turn, tracing the frontal positions and the lowest encircling isobar of the lows (for each day) onto the blank map. Note the changes in value of the central pressure of the low. Next, draw arrows connecting the centers of the lows for each day. Note that the connecting arrows become a little shorter each 24-hour period (the low is decelerating) and that the arrows point a little more toward the north each day (the system is curving toward the north in its trajectory).

To predict the Friday position of the center of the low, continue the smooth curve (which connects the central points of the low on Monday through Thursday) toward the northeast, curving it slightly more to the north. During the past three days, the speed of forward movement of the low has

decelerated by five knots during each 24-hour period. Consequently, a continued deceleration of five knots should be forecast for the next 24-hour period. Thus, the average speed of movement (forward) of the low over the next 24-hour period should be 15 knots. This means that the low will travel a distance of 360 nautical miles between Thursday and Friday. Place a mark along the smooth curve (trajectory of the low) previously drawn, at a distance of 360 nautical miles from the center of the low on Thursday. This is your predicted position of the center of the low on Friday—24 hours into the future. Also, since the central pressure of the low dropped by eight millibars in the past 24 hours, it can be expected to drop another eight millibars in the next 24 hours, and a central pressure of 984 millibars should be predicted for Friday.

Next, in exactly the same manner, draw 24-hour movement arrows for both the cold and the warm fronts. Also, be sure to draw in the precipitation shield, noting how it changes each 24-hour period. Do not hesitate to draw several arrows along each front in order to obtain the changes (if any) in orientation. Finally, connect the frontal-arrow tips for each front to smooth in the frontal positions. This is your forecast map for tomorrow (Friday).

Having gone through this drill systematically and carefully, your work chart and prognostic weather map should resemble very closely figure 12–13. If this is not the case, you'd better try again.

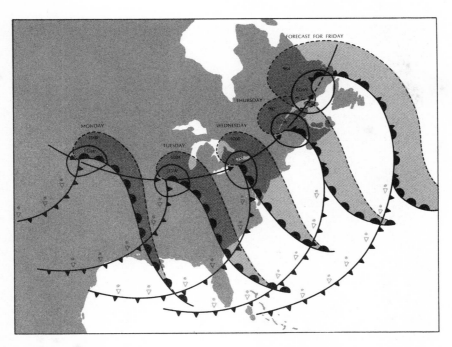

Figure 12–13.   Work chart and prognostic surface weather map for Friday

USING UPPER-AIR PATTERNS IN FORECASTING

Figures 12–14 and 12–15 are the ESSA Weather Bureau 0700 EST 500-Millibar Charts (average height 18,280 feet) for 13 and 14 December 1968, respectively. The chart notations are as explained in the previous section of this chapter. Note that on December 13th, a closed low-pressure cell (called a *cut-off low*) existed at this pressure level over southeastern Minnesota. On the following day, December 14th, the cut-off low had disappeared, and only a deep trough of low pressure, oriented NNE-SSW, extended from extreme northern Labrador to eastern Louisiana. These two types of upper-air patterns are most important to remember, as we shall see shortly.

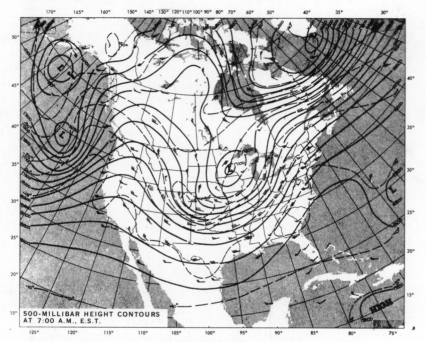

Figure 12–14.   ESSA 500-millibar chart for 13 December

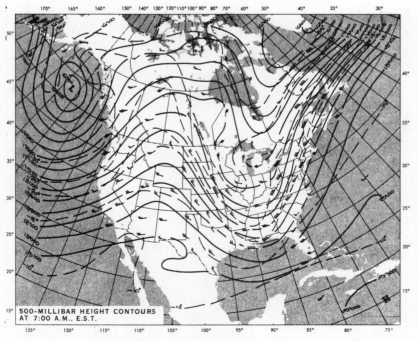

Figure 12–15.   ESSA 500-millibar chart for 14 December

Although storms usually follow the storm tracks shown in chapter 8, they are sometimes thrown off course by patterns of the air flow in the upper levels of the atmosphere. Figures 12–16 and 12–17 show how this happens. In both figures, the solid arrows show the contours, or flow pattern, at the 500-millibar level. The surface fronts (for the same time as the upper-air chart) are shown by conventional symbols. The shaded area on each chart shows where the warm sector (the area between the cold front and the warm front) will be 24 hours in the future.

On figure 12–15 note the cut-off low at the 500-millibar level just to the northwest of the surface frontal system. The crest of the surface frontal wave is just to the southeast of the center of the low at

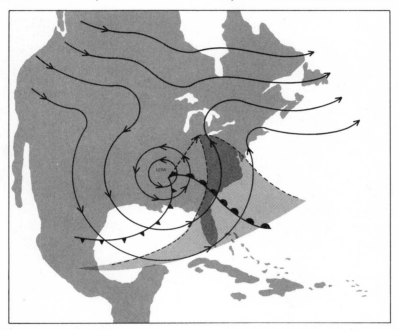

Figure 12–16. Surface pressure and frontal systems move slowly with this type of upper-air flow pattern.

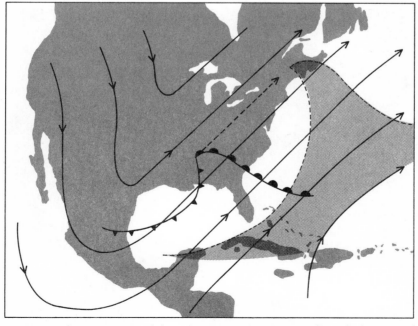

Figure 12–17. Surface pressure and frontal systems move very rapidly with this type of upper-air flow pattern.

500 millibars. This type of upper-air pattern, where a cut-off low is either directly over or at a small horizontal distance from a surface frontal wave, or low, causes the surface system to move quite slowly, perhaps intensify, or both. Note that in this case, the 24-hour movement of the surface system (represented by the shaded area) has been only 480 nautical miles.

Figure 12–17 illustrates how differently a surface frontal system, or low, will move if the upper-air pattern does *not* contain a cut-off low over, or at a short distance horizontally from, the surface system. The surface situation of figure 12–17 is identical to that of figure 12–16; however, the future movement of the surface system in figure 12–17 will be greatly different from that of the previous situation. In this case, where there is *not* a cut-off low aloft, the surface system will travel very rapidly toward the northeast under the strong steering current (from the southwest toward the northeast) at the 500-millibar level. In this case, the surface system will travel a distance of 1200 nautical miles in a 24-hour period, as compared to 480 nautical miles of the previous situation. Thus, it is readily seen that a knowledge and understanding of upper-air flow patterns is essential for accurate surface weather forecasting.

WEATHER SIGNS AND SOME GENERAL RULES FOR FORECASTING

Certainly, one should not hope to be a weather expert and infallible forecaster after reading a few books and articles on the subject and preparing a few prognostic charts. Meteorologists who have been involved in all aspects of the weather business for decades are the first to admit that they *still* learn something new about the weather with each day that passes.

There are many proverbs, rules, slogans, and signs pertaining to weather forecasting. Some are complementary, some are contradictory, and some are even correct—at least part of the time. Realizing this, we should heed carefully the counsel given over 2,300 years ago by the Greek poet, Aratus, who wrote in his *On Weather Signs:* "Make light of none of the warnings. It is a good rule to look for sign confirming sign. When two point the same way, forecast with hope. When three point the same way, forecast with confidence."

Weather signs *do* have prediction value if you have some sort of weather map available to you, and if you understand the atmospheric conditions which the signs indicate. If you can explain to yourself and to others—and you *should* be able to do so now—what the following weather signs and rules-of-thumb mean (in more-or-less scientific terminology), then they will be of definite use to you. The weather signs and forecasting rules already discussed in previous chapters will not be repeated here.

Signs

Fair weather will generally continue when:

> Summer fog clears off before noon.
> Cloud bases along mountains increase in height.
> Clouds tend to decrease in number.
> The wind blows gently from west to northwest.
> The temperature is "normal" for the time of year.
> The barometer is steady or rising slowly.
> The setting sun looks like a "ball of fire" and the sky is clear.
> The moon shines brightly and the wind is light.
> There is a heavy dew or frost at night.

Weather will generally change for the worse when:

> Cirrus clouds change to cirrostratus and lower and thicken.
> Rapidly moving clouds increase in number and lower in height.
> Clouds move in different directions at different heights.
> Clouds are moving from between NNE through east to south, and the wind speed increases with time.
> Altocumulus or altostratus clouds darken the western horizon and the barometer begins to fall rapidly.
> The wind shifts to the south or east. The greatest change occurs when the wind shifts from north through east to south.
> The wind blows strongly in the early morning.
> The temperature rises abnormally in the winter.
> The temperature is far above or below "normal" for that time of year.
> The barometer falls steadily.
> There is a downpour at night.
> A cold front, warm front, or occluded front approaches.

Weather will generally clear when:

> Cloud bases increase in height.
> A cloudy sky shows signs of clearing.
> The wind shifts to a westerly direction. The greatest change occurs when the wind shifts from east through south to west.
> The barometer rises rapidly.
> A cold front has passed 3 to 6 hours ago.

Rain or snow will generally occur:

When a cold, warm, or occluded front approaches.

In about 20 to 40 hours after the first cirrus-type clouds are noted to thicken and lower.

In about 14 to 26 hours after cirrostratus clouds are noted and there is a halo around the sun or moon.

Within about six to eight hours when the morning temperature is unusually high, the air is humid, and cumulus clouds are observed to be building.

Within about an hour in the afternoon when there is static on the radio and swelling cumulus clouds are observed.

When the sky is dark and threatening to the west.

When a southerly wind increases in speed and the clouds above are moving from the west.

When the wind—especially a north wind—backs (shifts in a counterclockwise manner from north to west to south).

When the barometer falls steadily.

Temperature will generally fall when:

The wind shifts into the north or northwest.
The wind continues to blow from the north or northwest.

The night sky is clear and the wind is light.
The barometer rises steadily in winter.
A cold front has passed.

Temperature will generally rise when:

The sky is overcast and there is a moderate southerly wind, at night.
The sky is clear during the day and there is a moderate southerly wind.
The wind shifts from the west or northwest to south.
A warm front has passed.

Fog will generally form when:

The sky is clear at sunset, the wind is light, and the air is humid.
Warm rain is falling through cold air ahead of a warm front.
There is a large temperature difference between relatively warm water and much colder air above it.
There is a sustained flow of warm, moist air northward (from the south) over a colder surface (either land or water).
Remember to watch carefully the temperature-dew point spread.

Rules of Thumb

Wind Direction	Sea-Level Pressure Millibars (Inches)	General Forecast
SW to NW	1019.3 (30.10) to 1022.7 (30.20) and steady	Fair, with little temperature change, for 1 to 2 days.
SW to NW	1019.3 (30.10) to 1022.7 (30.20) rising rapidly	Fair, followed within 2 days by rain.
SW to NW	1022.7 (30.20) or higher and steady	Continued fair with little temperature change.
SW to NW	1022.7 (30.20) or higher falling slowly	Fair for 2 days with slowly rising temperature.
S to SE	1019.3 (30.10) to 1022.7 (30.20) falling slowly	Rain within 24 hours.
S to SE	1019.3 (30.10) to 1022.7 (30.20) falling rapidly	Increasing winds and rain within 12 to 24 hours.
SE to NE	1019.3 (30.10) to 1022.7 (30.20) falling slowly	Increasing winds and rain within 12 to 18 hours.
SE to NE	1019.3 (30.10) to 1022.7 (30.20) falling rapidly	Increasing winds and rain within 12 hours.
SE to NE	1015.9 (30.00) or below falling slowly	Rain will continue 1 to 3 days, perhaps even longer.
SE to NE	1015.9 (30.00) or below falling rapidly	Rain with high winds in a few hours. Clearing within 36 hours—becoming colder in winter.
E to NE	1019.3 (30.10) or higher falling slowly	In summer, with light winds, rain may not fall for 2 to 3 days. In winter, rain within 24 hours.
E to NE	1019.3 (30.10) or higher falling rapidly	In summer, rain probably within 12 to 24 hours. In winter, rain or snow within 12 hours and increasing winds.

Rules of Thumb (continued)

Wind Direction	Sea-Level Pressure Millibars (Inches)	General Forecast	Wind Direction	Sea-Level Pressure Millibars (Inches)	General Forecast
S to SW	1015.9 (30.00) or below rising slowly	Clearing within a few hours. Then fair for several days.	E to N	1009.1 (29.80) or below falling rapidly	Severe storm (typical Nor'easter) in a few hours. Heavy rains or snowstorm. Followed by a cold wave in winter.
S to E	1009.1 (29.80) or below falling rapidly	Severe storm within a few hours. Then clearing within 24 hours—followed by colder in winter.	Hauling to W	1009.1 (29.80) or below rising rapidly	End of the storm. Followed by clearing and colder.

CONCLUSION

In this modern era of power-driven vessels with radio, radar, electronic aids, and satellite communications, there is an inevitable tendency for professional mariners and weekend sailors to be somewhat less weatherwise than their predecessors in square-riggers. But the practical value to all seamen of knowing something about the weather and the sea has not lessened with the passage of time. In fact, in this scientific, technological, and competitive age, the majority of seafarers agree that it has *increased*. The wind and the sea are still as powerful as ever. Hurricanes, typhoons, fog, ice, and adverse currents, are still factors to be respected and avoided. Many a modern maritime casualty has had weather and sea conditions or both as a contributory cause.

Appendix A
HANDY CONVERSION TABLES

Table A–1   Meters to Feet

Meters	Feet	Meters	Feet	Meters	Feet
1	3.3	10	32.8	100	328.1
2	6.6	20	65.6	200	656.2
3	9.8	30	98.4	300	984.3
4	13.1	40	131.2	400	1312.3
5	16.4	50	164.0	500	1640.4
6	19.7	60	196.9	600	1968.5
7	23.0	70	229.7	700	2296.6
8	26.2	80	262.5	800	2624.7
9	29.5	90	295.3	900	2952.8
				1000	3280.8

Table A–2   Nautical Miles to Statute Miles and Kilometers

Nautical Miles	Statute Miles	Kilometers	Nautical Miles	Statute Miles	Kilometers
1	1.15	1.85	10	11.52	18.53
2	2.30	3.71	20	23.03	37.07
3	3.46	5.56	30	34.56	55.60
4	4.61	7.41	40	46.06	74.13
5	5.76	9.27	50	57.58	92.66
6	6.91	11.12	60	69.09	111.20
7	8.06	12.97	70	80.61	129.73
8	9.21	14.83	80	92.12	148.26
9	10.36	16.68	90	103.64	166.79

NOTE: For larger distances, merely shift the decimal point.

Table A–3   Knots to Miles Per Hour and Meters Per Second

Knots	Miles Per Hour	Meters Per Second	Knots	Miles Per Hour	Meters Per Second	Knots	Miles Per Hour	Meters Per Second
1	1.2	0.5	10	11.5	5.1	110	126.7	56.6
2	2.3	1.0	20	23.0	10.3	120	138.2	61.8
3	3.5	1.5	30	34.5	15.4	130	149.7	66.9
4	4.6	2.1	40	46.1	20.6	140	161.2	72.1
5	5.8	2.6	50	57.6	25.7	150	172.7	77.2
6	6.9	3.1	60	69.1	30.9	160	184.2	82.4
7	8.1	3.6	70	80.6	36.0	170	195.8	87.5
8	9.2	4.1	80	92.1	41.2	180	207.3	92.7
9	10.4	4.6	90	103.6	46.3	190	218.8	97.8
			100	115.2	51.5	200	230.3	103.0

Table A–4   Speed and Distance Table

Knots	Nautical Miles Per Day	Nautical Miles Per Week	Knots	Nautical Miles Per Day	Nautical Miles Per Week	Knots	Nautical Miles Per Day	Nautical Miles Per Week
3	72	504	10.5	252	1,764	18	432	3,024
3.5	84	588	11	264	1,848	18.5	444	3,108
4	96	672	11.5	276	1,932	19	456	3,192
4.5	108	756	12	288	2,016	19.5	468	3,276
5	120	840	12.5	300	2,100	20	480	3,360
5.5	132	924	13	312	2,184	20.5	492	3,444
6	144	1,008	13.5	324	2,268	21	504	3,528
6.5	156	1,092	14	336	2,352	21.5	516	3,612
7	168	1,176	14.5	348	2,436	22	528	3,696
7.5	180	1,260	15	360	2,520	22.5	540	3,780
8	192	1,344	15.5	372	2,604	23	552	3,864
8.5	204	1,428	16	384	2,688	23.5	564	3,948
9	216	1,512	16.5	396	2,772	24	576	4,032
9.5	228	1,596	17	408	2,856	24.5	588	4,116
10	240	1,680	17.5	420	2,940	25	600	4,200

Table A–5   Visible Distance to the Horizon
(Based on Height of Observer's Eye)

Height (Feet)	Visible Distance			Dip of Horizon (Min./Sec.)	Height (Feet)	Visible Distance			Dip of Horizon (Min./Sec.)
	Nautical Miles	Statute Miles	Kilometers			Nautical Miles	Statute Miles	Kilometers	
5	2.57	2.96	4.77	2' 10"	70	9.61	11.06	17.81	8 06
10	3.63	4.18	6.73	3 04	100	11.48	13.22	21.28	9 41
15	4.45	5.12	8.24	3 45	150	14.06	16.19	26.07	11 52
20	5.13	5.92	9.53	4 20	200	16.23	18.69	30.09	13 42
25	5.74	6.61	10.64	4 51	300	19.88	22.90	36.87	16 44
30	6.29	7.24	11.65	5 18	400	22.96	26.44	42.57	19 32
40	7.26	8.36	13.46	6 07	500	25.67	29.56	47.59	21 39
50	8.12	9.35	15.05	6 51	1,000	36.30	41.80	67.30	30 37

Table A–6   Altitude and Atmosphere Table

Height (Thousands of feet)	Pressure		Density (x 10⁴) Lbs. Per Cu. Ft.	Temperature (Degrees C.)
	Millibars	Pounds Per Sq. Ft.		
0	1,013	2,116	765	+15.0
1	977	2,040	743	+12.5
2	942	1,967	721	+11.0
3	908	1,897	700	+ 8.5
4	875	1,828	679	+ 7.0
5	843	1,760	659	+ 5.1
6	812	1,696	639	+ 3.1
7	781	1,632	620	+ 1.1
8	752	1,572	601	− 0.9
9	724	1,513	583	− 2.8
10	697	1,455	565	− 4.8
12	644	1,346	530	− 8.8
14	595	1,243	497	−12.7
16	549	1,147	466	−16.7
18	506	1,057	436	−20.6
20	466	972	407	−22.4
25	376	785	343	−34.5
30	301	628	286	−44.4
35	238	498	237	−54.3
40	188	392	188	−56.5
45	147	308	148	−56.5
50	116	242	116	−56.5
55	91	196	92	−56.5
60	72	150	72	−56.5
65	59	118	57	−56.5
80	27.5	57	27.6	−46.3
100	10.8	22.6	10.1	−40.3
120	4.6	9.6	3.96	−21.6
140	2.06	4.3	1.66	− 3.3
160	0.97	2.0	0.75	+ 9.7
180	0.46	0.97	0.36	+ 3.1
200	0.21	0.44	0.18	−18.2

Table A–7   Temperatures: Degrees Centigrade to Fahrenheit

Degrees Centigrade	0	1	2	3	4	5	6	7	8	9
+40	104.0	105.8	107.6	109.4	111.2	113.0	114.8	116.6	118.4	120.2
+30	86.0	87.8	89.6	91.4	93.2	95.0	96.8	98.6	100.4	102.2
+20	68.0	69.8	71.6	73.4	75.2	77.0	78.8	80.6	82.4	84.2
+10	50.0	51.8	53.6	55.4	57.2	59.0	60.8	62.6	64.4	66.2
+ 0	32.0	33.8	35.6	37.4	39.2	41.0	42.8	44.6	46.4	48.2
− 0	32.0	30.2	28.4	26.6	24.8	23.0	21.2	19.4	17.6	15.8
−10	14.0	12.2	10.4	8.6	6.8	5.0	3.2	1.4	− 0.4	− 2.2
−20	− 4.0	− 5.8	− 7.6	− 9.4	−11.2	−13.0	−14.8	−16.6	−18.4	−20.2
−30	−22.0	−23.8	−25.6	−27.4	−29.2	−31.0	−32.8	−34.6	−36.4	−38.2
−40	−40.0	−41.8	−43.6	−45.4	−47.2	−49.0	−50.8	−52.6	−54.4	−56.2

Note: The vertical axis lists degrees Centigrade in ten-degree increments; the horizontal axis lists degrees Centigrade in one-degree increments. For example, to convert −23 degrees Centigrade to Fahrenheit, read down the vertical axis to −20 and across the horizontal axis to 3. The answer in this case is −9.4 degrees Fahrenheit.

Table A–8   Temperature Differences: Degrees Centigrade to Fahrenheit

Degrees Centigrade	Degrees Fahrenheit	Degrees Centigrade	Degrees Fahrenheit
1	1.8	6	10.8
2	3.6	7	12.6
3	5.4	8	14.4
4	7.2	9	16.2
5	9.0	10	18.0

Note: This table indicates the number of degrees Fahrenheit the temperature will rise for each increase in degrees Centigrade. For instance, if the temperature increases 7 degrees Centigrade, the comparable temperature in degrees Fahrenheit will increase 12.6 degrees. (A temperature increase from 10°C to 17°C equals an increase from 50° to 62.6°F.)

Table A–9(a)   Inches of Mercury (in round numbers) to Millibars

Inches of Mercury	Millibars	Inches of Mercury	Millibars	Inches of Mercury	Millibars
1	33.86	11	372.50	21	711.14
2	67.73	12	406.37	22	745.01
3	101.59	13	440.23	23	778.87
4	135.46	14	474.09	24	812.73
5	169.32	15	507.96	25	846.60
6	203.18	16	541.82	26	880.46
7	237.05	17	575.69	27	914.33
8	270.91	18	609.55	28	948.19
9	304.78	19	643.41	29	982.05
10	338.64	20	677.28	30	1015.92
				31	1049.78
				32	1083.65

Table A-9(b)   Inches of Mercury (in tenths and hundredths) to Millibars

Inches of Mercury	0.00	.01	.02	.03	.04	.05	.06	.07	.08	.09
0.0	0.00	0.34	0.68	1.02	1.35	1.69	2.03	2.37	2.71	3.05
0.1	3.39	3.73	4.06	4.40	4.74	5.08	5.42	5.76	6.10	6.43
0.2	6.77	7.11	7.45	7.79	8.13	8.47	8.80	9.14	9.48	9.82
0.3	10.16	10.50	10.84	11.18	11.51	11.85	12.19	12.53	12.87	13.21
0.4	13.55	13.88	14.22	14.56	14.90	15.24	15.58	15.92	16.25	16.59
0.5	16.93	17.27	17.61	17.95	18.29	18.63	18.96	19.30	19.64	19.98
0.6	20.32	20.66	21.00	21.33	21.67	22.01	22.35	22.69	23.03	23.37
0.7	23.70	24.04	24.38	24.72	25.06	25.40	25.74	26.08	26.41	26.75
0.8	27.09	27.43	27.77	28.11	28.45	28.78	29.12	29.46	29.80	30.14
0.9	30.48	30.82	31.15	31.49	31.83	32.17	32.51	32.85	33.19	33.53

Table A-9(a) converts inches of mercury in round numbers to millibars. Table A-9(b) converts inches of mercury in tenths and hundredths to millibars. Both tables can be used together to solve a conversion problem.

Example: Convert 29.73 inches of mercury to millibars.
Solution: First determine from table A-9(a) the number of millibars in 29 inches of mercury.

Then determine from table A-9(b) the number of millibars in .73 inches of mercury. In table A-9(b), the vertical axis lists inches of mercury in tenths of an inch. The horizontal axis lists inches of mercury in hundredths of an inch.

To convert .73 inches, read down the vertical axis to .7 and across the horizontal axis to .03.

Now add the data from both tables for the answer.

From table A-9(a)	982.05 mbs
From table A-9(b)	24.72 mbs
Answer:	1006.72 mbs

Appendix B
RECOMMENDED BOOKS AND PERIODICALS

BOOKS

Conant, James B. *On Understanding Science.* New York: The New American Library, 1951.

Douglas, Marjory S. *Hurricane.* New York: Rinehart and Co., 1958.

Dunn, G. E. and Miller, B. I. *Atlantic Hurricanes.* Baton Rouge: Louisana State University Press, 1960.

Harding, E. T. and Kotsch, W. J. *Heavy Weather Guide.* Annapolis: U. S. Naval Institute, 1965.

Hess, S. L. *Introduction to Theoretical Meteorology.* New York: Holt, Rinehart and Winston, Inc., 1959.

Huschke, R. E., ed. *Glossary of Meteorology.* Boston: American Meteorological Society, 1959.

Kendrew, W. G. *The Climates of the Continents,* 5th ed. Fair Lawn, N. J.: Oxford University Press, 1961.

Meteorology for Mariners, Met. O. 593, 2nd ed. London: Her Majesty's Stationery Office, 1967.

Middleton, W.E.K. and Spilhaus, A.F. *Meteorological Instruments,* 3rd rev. ed. Toronto: University of Toronto Press, 1953.

Neuberger, H. *Introduction to Physical Meteorology.* University Park: The Pennsylvania State University Press, 1951.

Neuberger, H. and Stevens, F. B. *Weather and Man.* Englewood Cliffs: Prentice-Hall, Inc., 1948.

Noel, Capt. J. V., Jr., ed. *Knight's Modern Seamanship.* New York: D. Van Nostrand, 1966.

Panofsky, H. *Introduction to Dynamic Meteorology.* University Park: The Pennsylvania State University Press, 1956.

Reiter, E. R. *Jet Streams.* Garden City: Doubleday and Co., Inc., 1967.

Riehl, H. *Introduction to the Atmosphere.* New York: McGraw-Hill Book Co., Inc., 1965.

Smithsonian Meteorological Tables. Washington: The Smithsonian Institution, 1951.

Stewart, G. R. *Storm.* New York: Modern Library, Inc. 1947.

Sverdrup, H. U. *Oceanography for Meteorologists.* Englewood Cliffs: Prentice-Hall, Inc., 1942.

Trewartha, G. T. *An Introduction to Climate,* 3rd ed. New York: McGraw-Hill Book Co., Inc., 1954.

U. S. Navy. *Marine Climatic Atlas of the World,* volumes on the Atlantic, Pacific, and Indian Oceans. Washington: Government Printing Office, published beginning 1955.

Willett, H. C. and Sanders, F. *Descriptive Meteorology,* 2nd ed. New York: Academic Press, Inc., 1959.

PERIODICALS

Average Monthly Weather Resume and Outlook. Semi-monthly. Washington: ESSA Weather Bureau.

Bulletin of the American Meteorological Society. Monthly. Boston: American Meteorological Society.

Bulletin of the World Meteorological Organization. Quarterly. Geneva, Switzerland: World Meteorological Society.

Journal of Applied Meteorology. Monthly. Boston: American Meteorological Society.

Mariner's Weather Log. Monthly. Washington: ESSA Weather Bureau.

Monthly Weather Review. Monthly. Washington: Government Printing Office.

Weatherwise. Bimonthly. Boston: American Meteorological Society.

Index

*The text of this book is set in ten point Linotype Caledonia
with two points of leading with display in Optima.*

*The entire book was composed, printed and bound by
The Kingsport Press, Kingsport, Tennessee. Text stock is 60#
Clear Spring Book Antique Offset White by Westvaco and
Cover Material is Holliston's White Lexotone, 17 point.*

*Designed by Harvey Satenstein.
Editorial production by Peter H. Spectre.
Line drawings by William J. Clipson.*